K3585 .O47 2003
0134110643894
Olsen, Marco A.

Analysis of
Stockholm C
c2003.

ANALYSIS OF THE STOCKHOLM CONVENTION ON PERSISTENT ORGANIC POLLUTANTS

MARCO A. OLSEN

Federal University of Espirito Santo UFES

OCEANA PUBLICATIONS, INC.
DOBBS FERRY, NY

Information contained in this work has been obtained by Ocean Publications from sources believed to be reliable. However, neither the Publisher nor its authors guarantee the accuracy or completeness of any information published herein, and neither Ocean nor its authors shall be responsible for any errors, omissions or damages arising from the use of this information. This work is published with the understanding that Ocean and its authors are supplying information, but are not attempting to render legal or other professional services. If such services are required, the assistance of an appropriate professional should be sought.

You may order this or any Oceana publication by visiting Oceana's website at http://www.oceanalaw.com.

Library of Congress Control Number: 2003104604

ISBN 0-379-21506-3

© 2003 by Oceana Publications, Inc.

All rights reserved. No part of this publication may be reproduced or transmitted in any way or by any means, electronic or mechanical, including photocopy, recording, xerography, or any information storage and retrieval system, without permission in writing from the publisher.

Manufactured in the United States of America on acid-free paper.

SUMMARY TABLE OF CONTENTS

FOREWORD vii

ACKNOWLEDGMENTS ix

INTRODUCTION. xi

CHAPTER 1. FUNDAMENTALS OF PERSISTENT ORGANIC POLLUTANTS. 1

CHAPTER 2. SUBSTANCE PROFILE OF PERSISTENT ORGANIC POLLUTANTS. 13

CHAPTER 3. BACKGROUND ON GLOBAL ACTION ON HAZARDOUS CHEMICAL SUBSTANCES PRIOR TO THE UNITED NATIONS STOCKHOLM CONVENTION ON PERSISTENT ORGANIC POLLUTANTS. 43

CHAPTER 4. THE FIRST GLOBAL TREATY ON PERSISTENT ORGANIC POLLUTANTS (POPS): NEGOTIATING THE STOCKHOLM CONVENTION (1995-2001) 77

CHAPTER 5. LEGAL ANALYSIS OF THE STOCKHOLM CONVENTION ON PERSISTENT ORGANIC POLLUTANTS. 107

CONCLUSION 121

APPENDIX. TEXT OF THE STOCKHOLM CONVENTION ON PERSISTENT ORGANIC POLLUTANTS 127

GLOSSARY OF ACRONYMS 165

GLOSSARY OF TERMS. 167

INDEX. 173

TABLE OF CONTENTS

FOREWORD vii

ACKNOWLEDGMENTS ix

INTRODUCTION xi

CHAPTER 1. FUNDAMENTALS OF PERSISTENT ORGANIC POLLUTANTS 1

   A. What are Persistent Organic Pollutants? 2

   B. Chemical Properties of Persistent Organic Pollutants 4

   C. Toxicology of Persistent Organic Pollutants 5

   D. Production, Uses and Sources of Persistent Organic Pollutants 6

   E. A World of Pesticides 8

CHAPTER 2. SUBSTANCE PROFILE OF PERSISTENT ORGANIC POLLUTANTS 13

   A. Pesticides 14

      1. Aldrin 14

      2. Dieldrin 17

      3. Endrin 18

      4. Chlordane 20

      5. Dichlorodiphenyltrichloroethane (DDT) 22

      6. Heptachlor 26

      7. Mirex 27

      8. Toxaphene 29

   B. Industrial Chemicals 30

      1. Polychlorinated Biphenyls—PCBs 30

      2. Hexachlorobenzene—HCB 33

   C. Unintended Produced By-products and Contaminants 34

      1. Polychlorinated Dibenzo—p—Dioxins and Furans 34

   D. Examples of the Effects of POPs on Health and the Environment 37

      1. Some Reported POPs Effects in the United States 38

      2. In the Arctic Region 39

CHAPTER 3. BACKGROUND ON GLOBAL ACTION ON HAZARDOUS CHEMICAL SUBSTANCES PRIOR TO THE UNITED NATIONS STOCKHOLM CONVENTION ON PERSISTENT ORGANIC POLLUTANTS 43

   A. Political Setting: The Context and National Legal Measures 43

      1. The Nature of the Problem: Growing Awareness 43

      2. National Regulation of Hazardous Chemicals 44

         2.1 Sweden 45

    2.2 Brazil 48

    2.3 Bilateral (U.S.-Canada) 49

B. "Soft Law" Measures to Regulate Hazardous Chemical Substances on the International Level 49

    1. The International Register of Potentially Toxic Chemicals—IRPTC (1976) 50

    2. The United Nations Food and Agriculture Organization International Code of Conduct on the Distribution and Use of Pesticides (1985) 52

    3. UNEP's London Guidelines for the Exchange of Information on Chemicals in International Trade (1987). 54

    4. The United Nations Conference on the Environment and Development (UNCED) and Agenda 21 (1992) 55

    5. The UNEP Code of Ethics on the International Trade in Chemicals (1994) 57

    6. International Organization for the Sound Management of Chemicals (IOMC) (1995). 59

C. Customary Law and Principle 21 61

D. "Hard Law" Measures to Regulate Hazardous Chemical Substances—The First Multilateral Treaties 63

    1. Basel Convention on the Control of Transboundary Movements of Hazardous Wastes and Their Disposal (1989) 63

    2. International Labour Organization's Chemicals Convention and Recommendation (1990) 65

    3. The Bamako Convention on the Ban of the Import into Africa and the Control of Transboundary Movement of Hazardous Wastes within Africa (1991) . . 66

    4. The Convention on Long-Range Transboundary Air Pollution (LRTAP) and its Protocol on Persistent Organic Pollutants (1998). 67

    5. The Convention on the Prior Informed Consent Procedures for Certain Hazardous Chemicals and Pesticides in International Trade (The Rotterdam Convention, 1998) 71

CHAPTER 4. THE FIRST GLOBAL TREATY ON PERSISTENT ORGANIC POLLUTANTS (POPS): NEGOTIATING THE STOCKHOLM CONVENTION (1995-2001) 77

A. Preparatory Work for the Intergovernmental Negotiating Committee on POPs . . 79

B. First INC Meeting—Montreal, 29 June-3 July 1998 83

    1. Main Issues Negotiated 84

C. First "Criteria Experts Group" Meeting—Bangkok, October 1998 86

D. Second INC Meeting—Nairobi, 25-29 January 1999 87

E. Second Criteria Experts Group—Vienna, 14-18 June 1999 90

    a) Article "F"—listing of substances in Annexes A, B or C. 91

    b) Article "O"—Conference of the Parties 94

    c) Annex "D"—Information Requirements and Criteria for the Proposal and Screening of Proposed Persistent Organic Pollutants 94

d) Annex "E"—Information Requirement for the Risk Profile 94

   e) Annex "F"—Information on Socio-Economic Considerations 94

 F. Third INC Meeting—Geneva, 6-11 September 1999 94

 G. Fourth INC Meeting—Bonn, 20-25 March 2000. 97

 H. Fifth INC Meeting—Johannesburg, 4-9 December 2001 98

 I. Conference of Plenipotentiaries on the Stockholm Convention on Persistent Organic Pollutants—Stockholm, 22-23 September 2001. 102

 J. The Role of Non-Governmental Organizations in Negotiating the Stockholm Convention 103

CHAPTER 5. LEGAL ANALYSIS OF THE STOCKHOLM CONVENTION ON PERSISTENT ORGANIC POLLUTANTS 107

 Preamble 108

 Article 1—Objective. 109

 Article 2—Definitions. 110

 Article 3—Measures to reduce or eliminate releases from intentional production and use 110

 Article 4—Register of Specific Exemptions. 111

 Article 5—Measures to reduce and eliminate releases from unintentional production 111

 Article 6—Measures to reduce or eliminate releases from stockpiles and wastes . . 112

 Article 7—Implementation Plans 112

 Article 8—Listing of Chemicals in Annexes A, B and C 113

 Article 9—Information Exchange 114

 Article 10—Public Information, Awareness and Education 114

 Article 11—Research, Development and Monitoring. 115

 Article 12—Technical Assistance 115

 Article 13—Financial Resources and Mechanisms 115

 Article 14—Interim Financial Arrangements 116

 Article 15—Reporting 116

 Article 16—Effectiveness Evaluation 116

 Article 17—Non-compliance 117

 Article 18—Settlement of Disputes. 117

 Article 19—Conference of the Parties 118

 Article 20 –Secretariat 118

 Article 21—Amendments to the Convention 118

 Article 22—Adoption and Amendments of Annexes 118

 Article 23—Right to Vote. 118

v

Article 24—Signature 118
Article 25—Ratification, Acceptance, Approval or Accession 119
Article 26—Entry into Force 119
Article 27—Reservations 119
Article 28—Withdrawal 119
Article 29—Depositary 119
Article 30—Authentic Texts 119
CONCLUSION 121
APPENDIX. TEXT OF THE STOCKHOLM CONVENTION ON PERSISTENT
ORGANIC POLLUTANTS 127
GLOSSARY OF ACRONYMS 165
GLOSSARY OF TERMS 167
INDEX 173

FOREWORD

Professor Marco Olsen has provided an exceptional service in writing this book. It provides the first scholarly examination of one of the most important environmental treaties of this new century. For the next several decades, the management and control of synthetic chemicals will be among the important challenges civilization confronts. Chemicals, far beyond their intended use, carry tremendous implications for impacts on the health of humans and other living beings. International control of synthetic chemicals is at its infancy, and the Stockholm Convention on Persistent Organic Pollutants provides the basic framework for global efforts to manage chemicals that do harm life.

While there is yet little international law on the subject of such chemicals, Sweden, through its able Chemicals Inspectorate, was able to reduce the unnecessary use of synthetic chemicals by over 75% from a baseline, without adversely impacting on the economy, and achieving measurable health benefits. Each of the twelve persistent organic pollutants banned under the Stockholm Convention was banned in the United States of America only after long and protracted litigation by manufacturers and applicators of those POPs. In the interim, a growing and significant record of avoidable injury to individuals resulted. Today, many, if not most, nations fall somewhere between the examples of Sweden and the U.S.A. Most nations still use some of these POPs, and do so without adequate regard for the collateral damage these chemicals are causing to their own citizens and to people and life elsewhere on the biosphere. There is ample evidence that continued use of these chemicals violates the general principle of international law framed in Principle 21 of the Stockholm Declaration on the Human Environment (1972), which posits each nation must assure that activities on its territory do not harm the environment of the territory of other nations or of the commons. Use of POPs demonstrably causes such harm.

While this treatise will assist many nations in understanding their obligations under the Stockholm Convention and facilitate their adherence to this new treaty, its most important service may be in providing an understanding about the legal framework for identifying new chemicals whose use must be curtailed or prohibited. Few of the over 60,000 synthetic chemicals in use today have been studied with respect to their unintended ambient environmental impacts. Laws like the Toxic Substances Control Act in the U.S.A. have not resulted in the hoped for careful scientific study of all these chemicals. There are chemicals that are close kin the twelve banned substances, and these relatives need to be studied with some urgency. The debates under the convention will concern which chemicals cause such collateral damage that it is not worth the risk to life to continue their use.

Foreword

Professor Olsen's contribution to our understanding of this new framework convention for global management of synthetic provides the essential foundation for the work of this coming century, which is securing environmental integrity for present and future generations.

White Plains, New York
Nicholas A. Robinson*

---

*   Gilbert & Sarah Kerlin Distinguished Professor of Environmental Law, Pace University School of Law, and Chair of the Commission on Environmental Law of the International Union for the Conservation of Nature and Natural Resources (IUCN).

ACKNOWLEDGMENTS

I am indebted to many people whose assistance, encouragement, insight and experience made this work possible. Most important, however, has been the support, encouragement, tolerance and consistently sound advice and expertise that I received from my Professor, guide and thesis advisor Nicholas Adams Robinson. Despite his incredibly demanding schedule and his worldwide travels in defense of the global environment, Professor Robinson was always ready to share with me his knowledge and time in discussing or reviewing early drafts of this thesis. For that I wish to express my deepest gratitude.

My understanding of the chemical implications and properties of persistent organic pollutants results from my association with a number of chemistry experts. Chief among them is Dr. David Rhani, whose experience, wisdom and critical suggestions significantly improved this work. Second, I was also gratified by the insights and recommendations of chemistry teaching assistant Che Wah Lee of Yale University, who offered advice on synthetic organic chemistry.

While I can take credit for the completion of this work, Dr. Catherine Tinker has provided critical guidance and accomplished the laborious task of editing this thesis. For her patience, caring, and sharp editorial judgment, I am forever grateful. I am also thankful to my colleagues Megan Brillault and Dru Stevenson for extraordinary editorial input.

I also wish to thank my professors Jeffrey Miller, Ann Powers and Richard Ottinger for their enthusiasm and encouragement.

Finally, I express my sincere gratitude to Pace University. This doctoral thesis was made possible by the financial support provided to me by of Pace University School of Law, without which I would never have been able to accomplish it. I owe a great debt to Pace University School of Law—Center for Environmental Legal Studies for helping me grow as a law professor through the collegial atmosphere of the Center for Environmental Legal Studies and the spirit of its founder Professor Nicholas Adams Robinson.

INTRODUCTION

The present work analyzes the Stockholm Convention on Persistent Organic Pollutants[1] (POPs) prepared under the auspices of the United Nations Environment Programme Chemicals Division. The treaty was adopted at the Conference of Plenipotentiaries in Stockholm on May 24th, 2001, and is still open for signature at United Nations Headquarters in New York until May 22nd, 2002. This treaty is the first international legal instrument to focus attention on the dangers of certain persistent organic pollutants, chemicals that are commonly used as pesticides in agriculture and to control insects causing diseases like malaria. At the same time, these chemical substances are carcinogenic, mutagenic and teratogenic, posing grave risks to human and animal health and the environment. The Stockholm Convention on persistent organic pollutants is a comprehensive global attempt to reduce the risks to human health and the environment from the release of certain persistent organic pollutants, currently known as "the dirty dozen."

Chapter One begins with an overview of the fundamentals of Persistent Organic Pollutants; specifically, their chemical properties, the toxicology of POPs and their common and/or historical uses. Chapter One also discusses the history of POPs contamination, how the modern era of synthetic pesticide production began and its effect on today's world.

Chapter Two concentrates on the science and chemistry of the twelve chemicals addressed in the Stockholm Convention. In additional, this chapter provides explicit examples of the harmful effects that these chemicals have on human populations.

Chapter Three examines the politics of "the dirty dozen" or POPs. In the 1960s, awareness from the "Green Revolution" drove governments to take action. Developed countries began passing national regulations; bilateral treaties were formed between countries such as the United States and Canada. The United Nations took steps, beginning in 1976, to address pollution by these chemicals. "Soft laws" were proposed by the United Nations Environment Programme, the World Health Organization, and the United Nations Food and Agriculture Organization as guidelines for industry. In 1989, the Basel Convention was the first international treaty addressing hazardous waste, though not specifically the twelve POPs chemicals.

Chapter Four begins discussing the intricacies of the Stockholm Convention. The international community began working on this treaty in 1995, holding numerous scientific and legal committee meetings. This chapter chronicles the

---

[1] The Stockholm Convention on Persistent Organic Pollutants (hereinafter Stockholm Convention), UNEP Doc. UNEP/POPS/CONF/2, May 22, 2001, *at* http://www.chem.unep.ch/sc/documents/meetings/dipcon/conf-2/eng/conf-2e.doc.

Stockholm Convention negotiations. Chapter Four also looks at the role of non-governmental organizations (NGOs) in negotiating the treaty.

Finally, Chapter Five analyzes the thirty articles of the Stockholm Convention. The goal of the convention—the reduction and/or elimination of certain chemical substances affecting human health and the environment—will be achieved through the implementation of the first three annexes when the identified chemicals are eliminated. However, one of the most important provisions of the treaty is found in Annex D, which allows the listing of new, additional chemicals not listed in Annexes A through C. This provision for adding new substances based on developing scientific data is an important feature of the new treaty.

When Persistent Organic Pollutants were first identified and their adverse effects were known by a few scientists, the public was yet unaware of the dangers ahead. What was known was that these chemicals were very effective as used, primarily as pesticides. Some were even praised as "miraculous" and one, DDT, was used to powder all soldiers at one point in time in the 1950s when they returned from foreign battlefields to discourage transmission of pests. As the scientific community realized their true danger, this knowledge was communicated to the rest of society.

The Stockholm Convention is a very important step towards phasing out these chemicals, attempting to curb their manufacture, transportation, and use to reduce their dangerous effect on humans, animals and the environment. The information contained in the following pages will allow readers to appreciate the steps that have been taken to reach the agreements made by the international community.

Hopefully, these pages will also give readers an appreciation of how much there is still to achieve to eliminate or control all persistent organic pollutants.

CHAPTER 1

FUNDAMENTALS OF PERSISTENT ORGANIC POLLUTANTS

Soon after World War II, societies faced the need to reconstruct cities, restructure communities, restore and/or build most public services, increase food supply, and improve public health, among other needs. A widespread occurrence of insect pests due to poor sanitary conditions prompted the chemical manufacturing industry to begin research and development to launch new and effective tools for dealing with these pests, hoping to save crops and eradicate disease-carrying insects.

Since then, it is estimated that more than 100,000 anthropogenic chemical substances have entered into the market, of which almost 75,000 are still being widely used by nations around the world.[1] Worst of all, most of these chemicals are not yet tested for their safety for use in human species, or in the biosphere.

In the United States alone, the annual production of synthetic organic chemical substances from 1950 to 1985 increased from 24 billion to 225 billion pounds.[2] They are used as food additives, medicines, agrochemicals, polymers and plastics, etc.

Agricultural policies in almost every country around the world have focused on output growth in order to provide sufficient food supply to the ever-increasing human population, currently at 6.2 billion and is projected to reach 10 to 14 billion by 2100 based on the most conservative projections.[3] The expansion of food production for this growing population has been associated with the use of certain chemical substances as fertilizers and pesticides at an alarming rate since the Second World War.[4] Clear warnings have been raised by the scientific community and environmentalists[5] about the problems caused by the use of such agents.

The use of chemical agents to control insects, pests, fungi, rodents, and pathogenic microorganisms, and to increase the fertility of soil certainly enhances productivity, but their overuse or misuse threatens the health of all living species, including humans. Continuous, long-term exposure to pesticides and their residues in our food, water, soil, and air is hazardous to humans and other species.

An epidemiological investigation conducted in 1983 concluded that approximately 10,000 people die each year in developing countries from pesticide poi-

---

1 J. Harte, Toxics A to Z: A Guide To Everyday Pollution Hazards 207-209 (University of California Press 1991).
2 G. Speth, *Environmental Pollution* Earth '88: Changing Geographic Perspectives, National Geographic Society. Washington, DC.
3 Earth Report 2000 (Ronald Bailey editor, 2000).
4 Ruth Norris et al., Pills, Pesticides & Profits 16 (1982).
5 Rachel Carson, Silent Spring, (Houghton Mifflin, 1962).

soning, and around 400,000 suffer acute health problems as a consequence of pesticide exposure.[6] What is worse is that the effects are usually not limited to the specific area where the pesticides are used, but rather are dispersed through water, air and soil and through the food chain.[7] Due to the fact that humans are at the top of the food chain, high concentrations of persistent organic pollutants are often accumulated and can be detected in human tissues, adipose tissue and milk. Almost 90% of the exposure to organochlorinated compounds among humans occurs via diet.[8]

A. What are Persistent Organic Pollutants?

One class of such toxic chemicals has been studied and classified by scientists as *Persistent Organic Pollutants*, (POPs). They are considered to be among the most dangerous compounds ever produced by chemical companies. Several organizational entities were created for international cooperation in studying and managing these persistent organic pollutants.

Acting on a recommendation made at the United Nations Conference on Environment and Development in 1992 in Rio de Janeiro, Brazil, the Inter-Organization Program for the Sound Management of Chemicals (IOMC) was established in 1995 by a number of participating international agencies with a specific objective to strengthen cooperation and increase international coordination in the field of chemical safety.[9]

In May 1995, the United Nations Environment Programme Governing Council adopted Decision 18/32 regarding Persistent Organic Pollutants, and determined that IOMC should work jointly with the International Program on Chemical Safety (IPC) and the Intergovernmental Forum on Chemical Safety (IFCS) to study and evaluate twelve chemical compounds initially identified on the basis of scientific consensus as to their hazardous nature.[10] The final assessment report assembled all existing information on the pertinent biochemistry and toxicology, transportation and disposal of these twelve chemical substances as well as the alternative products and costs to substitute them for the production and use of persistent organic pollutants.[11]

---

6   THE ORGANIC FOOD GUIDE (A. Gear ed., 1983).

7   STUART HARRAD, PERSISTENT ORGANIC POLLUTANTS: ENVIRONMENTAL BEHAVIOUR AND PATHWAYS OF HUMAN EXPOSURE 105-120 (2000).

8   Final Report, J. B.Greig and A.G. Smith, *Assessment of Early Signs of Biological Action Following Human Exposure to Polyhalogenated Dibenzo-p-dioxins and Related Substances*, (1998).

9   See Inter Organization Program for the Sound Management of Chemicals *at* http://www.who.int/iomc (last visited April 8, 2002).

10 L. RITTER ET AL., CANADIAN NETWORK OF TOXICOLOGY CENTERS, PERSISTENT ORGANIC POLLUTANTS: AN ASSESSMENT REPORT ON DDT, ALDRIN, DIELDRIN, ENDRIN, CHLORDANE, HEPTACHLOR, HEXACHLOROBENZENE, MIREX, TOXAPHENE, POLYCHLORINATED BIPHENYLS, DIOXINS, AND FURANS (1995).

11 *See id.*

These twelve, by and large chlorinated hydrocarbon compounds, can be grouped into three categories:

Pesticides: Aldrin, Chlordane, Dieldrin, DDT, Endrin, Heptachlor, Hexachlorobenzene, Mirex, and Toxaphene.

Industrial Chemicals: Polychlorinated biphenyls (PCBs).

Industrial by-products: Polychlorinated dibenzodioxins (Dioxins), and polychlorinated dibenzofurans (Furans).

Such chemical compounds, despite their immediate hazards when used as pesticides upon being once released into the environment, resist photolytic, biological, and chemical degradation, that is to say, they do not lose their effects and are long-lasting agents. The characteristic of **persistence** means the length of time a given chemical substance will remain in the environment before it is ultimately broken down or degraded into another and less hazardous substance.[12] More precisely speaking, every synthetic substance when released into the environment has a specific half-life, the time required for half such substance to be degraded. Such half-life for each of the above twelve compounds is generally in the order of hundreds of years.[13]

Generally, these chemicals are also **semi-volatile**, which means they evaporate slowly into the atmosphere.[14] In such circumstances, POPs enter the air and are able to travel considerably long distances on air currents, before eventually returning to the soil. In colder regions such as the Arctic, this volatility is reduced even further, resulting in accumulation greater than in other parts of the world.[15] Persistent Organic Pollutants are detected in areas such as the Arctic region, where these substances have never been used or produced and at threatening high levels that pose severe risks to the lives of both humans and wildlife.[16]

Another property of these pollutants is their extremely low solubility in water.[17] In other words, they do not dissolve considerably in water, but rather they dissolve readily in oil and fats, causing them to be known as substances of **high lipid solubility**. Since a large number of mammals found in the Arctic region possess high fat deposits, the concentration of persistent organic pollutants in fatty tissues of polar bears and sea lions, for example, is high and can increase by a factor of many thousands or millions as the chemicals move up the food chain.[18]

---

12 D. MACKAY, W.Y. SHIU & K.C. MA, ILLUSTRATED HANDBOOK OF PHYSICAL-CHEMICAL PROPERTIES AND ENVIRONMENTAL FATE FOR ORGANIC COMPOUNDS (Vol. I. Lewis Publishers 1992).
13 *Id*.
14 ROBERT L. LIPNICK, ET AL., PERSISTENT, BIOACCUMALATIVE, AND TOXIC CHEMICALS II: ASSESSMENT AND NEW CHEMICALS 14-20 (2000).
15 R. Freeman & J. Schroy, *Environmental Mobility of TCDD*, 14 CHEMOSPHERE 873 (1985).
16 L.A. Barrie et al., *Arctic Contaminants: Sources, Occurrence, and Pathways*, 1 SCI. TOTAL ENVT. 74 (1992).
17 *See id., supra* note 14.
18 *See id, supra* note 14.

Persistent organic pollutants can be transported over long distances through air, falling into remote regions away from their original point of emission.[19] These substances are also known for their bioaccumulation properties into the environment and other living species, including mankind.[20] Bioaccumulation, or **bioconcentration** is defined as the result of a balance between the rate of chemical uptake from water or air via the respiratory surface of the organism, gills and skin for instance, and the loss or elimination thereafter.[21] If the outflow is smaller than the inflow of a chemical, the balance is then bioaccumulated in that species.

Another characteristic of POPs is their **biomagnification** characteristic; this is the process where the chemical concentration of the substance in a given living organism achieves a level that exceeds that in the regular diet due to dietary absorption.[22] This in turn is uptaken by the next species in the food chain, thereby resulting in higher concentration of POPs.

According to scientific classification, any organic substances that present properties of being persistent, demonstrating bioaccumulation and also posing a threat to human health due to their toxicity, are described as Persistent, Bioaccumulative, Toxic Substance (PBTs).[23] PBTs pose serious concern at local, regional and global levels depending on the ways that such substances move around from one place to another.[24] Within PBTs appears the subclass of persistent organic pollutants, a group of chemical compounds that can be transported to long-range atmospheric distances and resulting in deposition.[25]

B. Chemical Properties of Persistent Organic Pollutants

A major class of persistent organic pollutants is called halogenated aromatic substances, which are most often chlorinated and originate almost entirely from anthropogenic activities associated largely with the manufacture, use and disposal of a number of chemical compounds.[26] Other substances, such as dioxins

---

19 D. Rappe et al., *Long Range Transport of PCDDs and PCDFs on Airborne Particles* 18 CHEMOSPHERE 1283-1290 (1989).

20 STUART HARRAD, PERSISTENT ORGANIC POLLUTANTS: ENVIRONMENTAL BEHAVIOUR AND PATHWAYS OF HUMAN EXPOSURE 145-155 (2000).

21 Gobas, FAPC & H.A. Morrison, *Bioconcentration & Bioaccumulation in the Aquatic Environment* in HANDBOOK FOR ENVIRONMENTAL PROPERTIES (R. Boethling & D. Mackay eds., 1999).

22 See id, supra note 19.

23 See id, supra note 14 at 1-14.

24 H.W. Walack et al., *Controlling Persistent Organic Pollutants—What Next?* 6 ENVTL. TOXICOLOGY PHARMACOLOGY 143 (1998).

25 H.W. Walack et al., *Controlling Persistent Organic Pollutants—What Next?* 6 ENVTL. TOXICOLOGY PHARMACOLOGY 143 (1998).

26 See id, supra note 14.

and furans, are chemicals that are produced in a wide range of manufacturing and combustion processes,[27] such as waste incineration.[28]

These chemicals are substances characterized by their resistance to degradation either by photolysis or by chemical or biological (aerobic or anaerobic) processes.[29] They bioaccumulate in the environment, in human tissues or in animal tissues without breaking down, posing serious threat to health in general due to their toxic properties.

The United Nations Environment Programme's Chemicals Division ("UNEP Chemicals")[30] recognizes over three hundred chemicals with properties that would categorize these substances as persistent organic pollutants. The implication of this recognition is that their uses should be halted and eliminated worldwide for the safeguarding of human and non-human life and the environment.

C. Toxicology of Persistent Organic Pollutants

Toxicology is the study of all adverse effects of chemical compounds on human or nonhuman health and the environment.[31] There is substantial toxicological evidence that there are numerous, mostly synthetic, chemical agents which are carcinogenic, mutagenic and/or teratogenic in humans. At the turn of this century, studies in humans showed that environmental and possibly internal chemical agents are causative factors in the development of cancer.[32] Two Japanese pathologists in 1915 described the first production of skin tumors in animals by the application of coal tar to the skin. Yamagawa and Ichikawa repeatedly applied crude coal tar to the ears of rabbits for a number of months, finally producing both benign and later malignant epidermal neoplasm (cancer).[33] During the last 150 years, a number of specific chemicals and chemical mixtures, industrial processes, and lifestyles have been casually related to increased incidents of a variety of human cancers and other adverse health effects.[34]

---

27   D. Crosby, K. Moilanen & A. Wong, *Environmental Generation and Degradation of Dibenzodioxins and Dibenzofurans* 5 ENVIRONMENTAL HEALTH PERSPECTIVE 259 (1973).

28   K. Olie, J. Lustenhouwer & O. Hutzinger, *Polychlorinated dibenzodioxins and Related Compounds in Incinerator Effluents* in CHLORINATED DIOXINS AND RELATED COMPOUNDS 227 (O. Hutzinger et al. eds., 1982).

29   See id. supra note 14.

30   UNEP Chemicals is the center for all chemicals-related activities of the United Nations Environment Programme. UNEP Chemical's Web site: http://www.chem.unep.ch/irptc/default.htm.

31   J. HARTE, TOXICS A TO Z: A GUIDE TO EVERYDAY POLLUTION HAZARDS 25-35 (University of California Press 1991).

32   M.B. SHIMKIN, CONTRARY TO NATURE (U.S. Department of Health, Education, and Welfare, Public Health Service, National Institutes of Health 1977); P.D. Lawley, *Historical Origins of Current Concepts of Carcinogenesis* 65 ADVANCED CANCER RESEARCH 17 (1994).

33   B. Allen, K. Crump & A. Shipp, *Correlation Between Carcinogenic Potency of Chemicals in Animals and Humans* 8 RISK ANALYSIS 531 (1998).

34   R. DOLL & R. PETTO, THE CAUSES OF CANCER (Oxford Univ. Press 1976).

It is a consensus among scientists that human exposure to persistent organic pollutants can cause the following effects:

a) Immune system changes;

b) Cancers and tumors at multiple sites;

c) Reproductive and sex-linked disorders;

d) Neurobehavioral impairment, such as learning disorders;

e) Increased incidence of diabetes;

f) Development of endometriosis.[35]

As far as wildlife exposure is concern, good evidence exists of the following kinds of injury:

a) Development of cancer;

b) Gross birth defects;

c) Reproductive failure and population decline;

d) Hormonal dysfunctions;

e) Feminization of males and masculinization of females;

f) Behavioral abnormalities; and

g) Deficiencies in the immune system.[36]

People are subject to POP contamination through their food or water supply. Exposure to sources of persistent organic pollutants can also pose serious threats through inhalation and contact with skin. The effects of POPs from food will accumulate in a mother's body and interfere with the development of the fetus, causing injury to the fetus that is only manifested when the child reaches puberty or adulthood.

D. Production, Uses and Sources of Persistent Organic Pollutants

Table 1: Initial World Production of Persistent Organic Pollutants[37]

Organochlorine Pesticides	Initial Producer	Year of first Production	Annual Global Use (Tones)
ALDRIN	SHELL	1948-1950	130,000
DIELDRIN	SHELL	1948-1950	130,000
ENDRIN	Velsicol and Shell		2,300

---

35   United Nations Development Program *Persistent Organic Pollutants (POPs) Resource Kit at* http://www.undp.org/get (2001). Endometriosis is a painful, chronic gynecological disorder in which uterine tissues grow outside the uterus.

36   United Nations Development Program *Persistent Organic Pollutants (POPs) Resource Kit at* http://www.undp.org/get (2001).

37   Source *at* http://www.greenpeace.org/%7Etoxics/tbg/tbg1a.html. Site visited on April 8, 2002.

Organochlorine Pesticides	Initial Producer	Year of first Production	Annual Global Use (Tones)
CHLORDANE	Velsicol	1947	150,000 (1950 to 1992)
DDT	Ciba-Geigy	1946	100,000
MIREX	Allied ChemicalCo Hooker Chemical	1946 1958-1959	Not available 1,500
HEPTACHLOR	Velsicol	1952	Not available
TOXAPHENE	Hercules Powder Co	1946	1,333,000
HCB		1945	100,000
PCB's	Monsanto	1929	1,200,000

Table 2: Producers of Persistent Organic Pollutants Worldwide[38]

Chemical	Market Name	Country of Origin	Company Producer
ALRIN	Aldrin Dispersivel Aldrex 5 Aldrex 600 Farmon Aldrin 30 Aldrin 30 Aldripoudre g Caritex	Germany Great Britain Netherlands USA Great Britain France Belgium	BASF AG ICI Ltd Shell Tenneco Inc. Tenants Consolidated Rohne-Poulenc SA Produits et Engrais Chimiquews du Portugal S.A.
DIELDRIN	Dieldrite	Netherlands Germany	Shell Merch, Sharp & Dohme
ENDRIN	Luxan endrin 20% Endrotox Agrine Endrex Endrex Endrex	Netherlands Austria USA Japan Japan Japan Japan Japan Japan	Cebeco-Handelsraad Kwizda Reichholt Chemicals Chugai Boeki Co. Ltd. Ishihara Sangyo Co. Takeda Chemicals Ind. Ltd Tomono Noyaku Sankei Chemicals Sumitomo Kagkku Kogyo
CHLORDANE	Chloroxone Gold Crest c-100 Octa-klor Termidid Velsicol1068	USA	Reichhold Chemicals Inc Sandoz AG
DDT	Dedelo Hildit	Great Britain India Spain Italy	ICI Ltd Hindustan Insecticides Ercross S.A. Ente Nazionale Idrocarburi
HEPTACHLOR	Heptachlor 40 ec Drinox h-34, Gold Crest h-60, Heptamul, Heptox, Termid	Germany	Hoechst AG Sandoz AG

---

38   WEBSITE *at* http://www.greenpeace.org/%7Etoxics/tbg/tbg6.html. Site visited on April 8, 2002.

Chemical	Market Name	Country of Origin	Company Producer
PCB's	Fenclor	Italy	Caffaro
	Pyrochlor	USA	Monsanto
	Pyrenol/Araclor	Great Britain	Monsanto
	Clophen	Germany	Bayer AG
	Inerteen	USA	Westinghous
	Kanechlor	Japan	Kanegafuchi Chemical Co.
	Pyralene	France	Prodelec
	Delor	Former Czechosl.	Chemco
DIOXINS, FURANS	Incinerators Chlorine Industries	Worldwide	Dow, Solvay, AKZO-Nobel, ICI
TOXAPHENE	Camphofene Camphoclor	Germany	Hoechst AG
	Strobane-t90	USA	Retzloff Chemical Co.
HEXACHLORO BENZENE	Tizoneb	Spain	Ercross S.A.
	Caritex	Belgium	Produits et Engrais Chimiquews du Portugal S.A.
	Triffor spuitzpulver	Austria	Kwizda
	Abavit universal	Germany	Schering AG
	trockenbeize	Japan	Hodogaya Chemical Co Ltd.

Source: WEBSITE at http://www.greenpeace.org/%7Etoxics/tbg/tbg6.html. Site visited on April 8, 2002.

E. A World of Pesticides

The term pest derives from the Latin *pestis,* which means plague. It is used to describe weeds, vertebrates, insects, mites, pathogens and other organisms that occur where humans do not want them. Due to an absence of natural enemies and changes in agricultural patterns, vast quantities of insects are able to reproduce at an alarming speed and dissipate entire populations typically causing serious agricultural problems typically in a short period of time.

"Pesticide" is a generic name for a variety of chemical agents that are classified more specifically on the basis of the pattern of use and organism killed. There are at times other more specific terms used, such as herbicide (weed control), insecticide, rodenticide, and fungicide; however, they are collectively referred to as pesticides.

Many ingenious methods have been invented in the attempt to control the invertebrates, vertebrates, and microorganisms that constantly threaten human's food supply and health. The use of a particular substance to control the spread of undesirable or harmful microorganisms dates back in history before 1000 B.C., when the Chinese used sulfur as a fumigant, and in the 1800s, when Europeans used it as a fungicide against powdery mildew on fruit.[39]

---

[39] THE SCIENCE OF LIFE: CONTRIBUTIONS OF BIOLOGY TO HUMAN WELFARE 243-294 (K.D. Fisher & A.U. Nixon, eds., 1972).

Burning sulfur yields sulfur dioxide which, when entering the respiratory system of a microorganism, produces sulfuric acid, thereby killing germs. Such a practice may still be used in certain rural regions of the world, particularly in developing countries.[40]

In the 16th century the Japanese used whale oil mixed with vinegar to spray on rice paddies and crop fields to prevent the development of insect larvae.[41] In the 17th century in Europe, water extracts of tobacco leaves were sprayed on plants as insecticides, and the seed of *Strychnos nuxvomica* (strychnine) was used to kill rodents, which then posed a serious human health threat in Europe and many other parts of the world.[42] In the 19th century, arsenic trioxide was used as a weed killer, primarily for dandelions. By the 1920s, the widespread use of arsenical pesticides in the United States and Western Europe caused considerable public concern because some treated fruits and vegetables were found to have toxic residues. Crop seeds pre-treated with such inorganic pesticides, provided by international food assistance programs to developing countries and in cases of emergencies, have mistakenly been processed as food, killing thousands in several incidents in the 20th century alone worldwide.[43]

The 1930s bloomed as the era of the modern synthetic chemical industry. This decade marked the development of a variety of pesticide agents, such as alkyl thiocyanate insecticides, dithiocarbamate fungicides, and ethylene dibromide, methyl bromide, ethylene oxide, and carbon disulfide as fumigants.[44] By the beginning of World War II there were a number of pesticides, including dichlorodiphenyltrichloroethane (DDT), under experimental investigation. A Nobel Prize was bestowed for its discovery.[45] In the post-war era, there was rapid development in the agrochemical field, with a plethora of insecticides, fungicides, herbicides, and other chemical agents introduced. No other field of synthetic organic chemistry has witnessed such a diversity of structures arising from the application of the principles of chemistry to the mechanisms of action in pests in order to develop selectivity and specificity in agents toward certain species, while reducing toxicity to other forms of life.

Paradoxically, there is no such a thing as a completely safe pesticide. There are, however, pesticides that can be used safely and that present a low level of risk to human health when applied with proper attention to the label's instructions. Despite the current conflagration over pesticide use and the presence of low

---

[40] NATIONAL RESEARCH COUNCIL: WASTE INCINERATION AND PUBLIC HEALTH (National Academy Press 2000).
[41] R. CREMLYN, PESTICIDES: PREPARATION AND MODE OF ACTION 125 (1978).
[42] PATRICK F. SCANLON, Nature and Use of Rodenticides, Piscicides, Avicides and Repellents (April 23, 2001).
[43] T. WILKINSON, SCIENCE UNDER SIEGE: THE POLITICIAN'S WAR ON NATURE AND TRUTH (Johnson Books 1998).
[44] R. CREMLYN, PESTICIDES: PREPARATION AND MODE OF ACTION (Wiley 1978).
[45] C. KIRBY, THE HORMONE WEEDKILLERS (BCPC 1980).

levels of residues in food, groundwater, soil, and air, these agents comprise integral components of our crop and health protection programs.

No one can doubt the efficacy of pesticides for the protection of crops in the field, thereby providing us with abundant, inexpensive, and ideally wholesome and attractive fruits and vegetables. The medical miracles accomplished by pesticides have been documented. To number but a few:

a) The suppression of a typhus epidemic in Naples, Italy, by DDT in the winter of 1943-1944;[46]

b) The control of river blindness (onchocerciasis) in West Africa by killing the insect vector (black fly) carrying the filarial for this disease with temephos (Abate);[47] and

c) The control of malaria in Africa, the Middle East, and Asia by eliminating the plasmodia-bearing mosquito populations with a variety of insecticides.[48]

The risk of their use, however, has been widely acknowledged. On a worldwide basis, intoxication attributed to pesticides has been estimated to be as high as three million cases of serious acute poisoning annually, with as many or more unreported cases and some 220,000 acute deaths, according to the World Health Organization's data.[49] Taking the United States as an example, in the state of California, some 25,000 cases of pesticide-related illness occur annually among agricultural workers.[50] That compares to the national estimate, which on the order of 80,000 cases per year.[51]

The incidence of poisoning is thirteen times higher in developing countries than in highly industrialized nations, which consume 85% of world pesticide production.[52] Recently published proceedings give a good overview of the dilemma in developing countries, where there are few regulations or enforcement measures controlling the registration and sale of pesticides.[53]

There is still a great need for advancement in disease vector control by pesticides. Some six hundred million people are at risk from schistosomiasis in the Middle East and Asia, and almost one billion people around the world harbor

---

[46] G.T. BROOKS, CHLORINATED INSECTICIDES, TECHNOLOGY AND APPLICATION 12-13 (1974).
[47] J. Walsh, *River Blindness: A Gamble Pays Off* 232 SCIENCE 922-925 (1986).
[48] F. MATSUMURA, TOXICOLOGY OF INSECTICIDES 122-28 (1985).
[49] WORLD HEALTH ORGANIZATION W.H.O: PUBLIC HEALTH IMPACT OF PESTICIDES USED IN AGRICULTURE (1990).
[50] G. Forget, *Pesticides: Necessary but Dangerous Poisons* 18 IDRC REP 4-5 (1989).
[51] M.J. Coye, J.A. Lowe, K.T. Maddy, *Biological Monitoring of Agricultural Workers Exposed to Pesticides: I. Cholinesterase Activity Determinators* 28 JOURNAL OF OCCUPATIONAL MEDICINE 619-627 (1986).
[52] G. Forget, *Pesticides: Necessary but Dangerous Poisons* 18 IDRC REP 4-5 (1989).
[53] INTERNATIONAL RESEARCH CENTER, IMPACT OF PESTICIDE USE ON HEALTH IN DEVELOPING COUNTRIES (G. Forget, T. Goodman & A. de Villiers eds., 1993).

pathological intestinal worm infestations.[54] The widespread use and misuse of the early, toxic pesticides created an awareness of the potential health hazards and the need to protect humankind and the environment from residues derived from their use. In the United States, it was not until 1906 that Dr. Harvey Wiley, the head of the Chemistry Division, U.S. Department of Agriculture urged federal legislation to make food adulteration a crime, creating the first Federal Food and Drugs Act (the "Wiley Act"),[55] which was later replaced in 1938 by the Federal Food, Drug and Cosmetic Act ("FDCA").[56] Specific pesticide amendments were passed in 1954 and 1958[57] requiring that pesticide tolerance levels were to be established for all agricultural commodities.[58] The 1958 amendment to FDCA's Delaney Clause (section 409),[59] stated "no additive shall be deemed safe if it is found to induce cancer when ingested by man or animal or, if it is found, after tests which are appropriate for the evaluation of the safety of food additives, to induce cancer in man or animals."[60]

Another important act passed by Congress in 1947 was the Federal Insecticide, Fungicide, and Rodenticide Act (FIFRA),[61] a labeling statute that would group all pest-control substances, initially only insecticides, fungicides, rodenticides, and herbicides. This law was to be administered by the United States Department of Agriculture (USDA). The primary goal of FIFRA is to provide federal control of pesticide distribution, sale, and use.[62] Later in 1961, Congress added nematicides, plant growth regulators, defoliants, and desiccants to the statute.[63] The authorization to deny, suspend, or cancel registration of products was granted to the USDA.[64] In 1971, this system was reorganized and the administrative authority was turned over to the newly created Environmental Protection Agency (EPA).[65] In 1976, Congress enacted the Toxic Substances Control Act

---

54 A. ALBERT, XENOBIOSIS, FOOD, DRUGS AND POISONS IN THE HUMAN BODY 113-116 (Chapman and Hall 1987).
55 Federal Food and Drugs Act of 1906, 34 STAT.768 (1906), *available at* http://www.fda.gov/opacom/laws/wileyact.htm (last visited Feb. 24, 2002).
56 21 U.S.C. § 329(a), *available at* http://www.fda.gov/opacom/laws/wileyact.htm (last visited Feb. 24, 2002).
57 Federal Food, Drug, and Cosmetic Act, Section 201.[321], *available at* http://www.uvm.edu/nusc/nusc237/201.html (last visited Feb. 24, 2002).
58 *See id.*
59 *See id.*
60 Report, Committee on Scientific and Regulatory Issues Underlying Pesticide Use Patterns and Agricultural Innovation, National Academy of Sciences, *Regulating Pesticides in Food: The Delaney Paradox*, (Washington, DC).
61 Federal Insecticide, Fungicide, and Rodenticide Act, 7 U.S.C. §§ 136-136y (2001).
62 *See id.*
63 *See id.*
64 PUBLIC POLICIES FOR ENVIRONMENTAL PROTECTION, HAZARDOUS WASTE AND TOXIC SUBSTANCE POLICIES, (Paul R. Portney & Robert N. Stavins eds., 2000).
65 *See id.*

(TSCA)[66] to give EPA the ability to track the 75,000 industrial chemicals currently produced and imported into the United States.[67] Under this statute, any chemical substance that poses risk of injury to human health or the environment is banned.[68] For example, section 2605(e) regulates disposal, labeling, processing, distribution and use of Polychlorinated biphenyls (PCBs).[69] Section 12(a) of TSCA allows EPA to restrict any chemical substance export in case of "an unreasonable risk of injury to health in within the United States or to the environment of the United States."[70] Section 12(b) demands exporters to notify EPA of their intent to export, and who then in turn notifies the importing country.[71] Additionally, the Consumer Product Safety Commission (CPSC), in the event of finding a chemical substance of "unreasonable risk of injury to consumers within the United States," has the authority to ban an export.[72]

---

66   The Toxic Substances Control Act, 15 U.S.C. §§ 2601-2692 (1976).
67   PUBLIC POLICIES FOR ENVIRONMENTAL PROTECTION, HAZARDOUS WASTE AND TOXIC SUBSTANCE POLICIES, (Paul R. Portney & Robert N. Stavins eds., 2000).
68   See id.
69   See 15 U.S.C § 2605(e) (1978).
70   15 U.S.C. § 2611(a)(2) (1992).
71   15 U.S.C. § 2611(b)(1) (1992).
72   15 U.S.C. § 2067(a)(1)(B) (1992).

CHAPTER 2

SUBSTANCE PROFILE OF PERSISTENT ORGANIC POLLUTANTS

Construction of international legal rules proceeds more efficiently when scientific consensus exists on the nature of the threat and the measures necessary to deal with it. All the findings of the scientific community need to be better understood and brought into the political decision-making process.[1] After years of experience in negotiating international treaties, UNEP has learned that political and economic opposition to significant changes can be disguised as a desire for more scientific information and research. The United Nations Conference on Environment and Development in 1992 in Rio de Janeiro adopted Agenda 21,[2] which calls for strengthening U.N. system to develop a stronger scientific basis for improving the environmental and developmental policy formulation.[3] Elsewhere Agenda 21 recognizes that:

> "to effectively integrate environmental and developmental in the policies and practices of each country, it is essential to develop and implement integrated, enforceable and effective laws and regulations that are based upon sound social, ecological, economic and scientific principles."[4]

Since the beginning of the negotiations of the Stockholm convention, negotiators felt more secure that their regulation of persistent organic pollutants would reflect, as much as possible, sound science. This would secure that the treaty stand the test of time and not be influenced by changing social and political conditions among nations. Science-based criteria, for making international environmental law, reflects awareness that science is rapidly advancing with regard to the factors that determine the behavior of chemicals in the environment and in organisms, including humans.[5] The comparatively rapid and vigorous response of international environmental law to the problem of toxic pollutants was due largely to the fact that scientific research results in this area were available early on and high quality, that researchers and governments were able to cooperate, and that policy makers were attentive to the scientist's description of the problems. Much of the science on persistent organic pollutants was used in preparing a number of existing environmental conventions and other instruments; in particular those listed below, have fully taken science into account:

a) Basel Convention on the Control of Transboundary Movements of Hazardous Wastes and their Disposal;

---

[1] Nicholas A. Robinson, *Legal Systems, Decision making, and the Science of Earth's Systems: Procedural Missing Links*, 27 ECOLOGY L.QTLY. 1077-1161 (2001).
[2] *See generally*, Agenda 21, June 13, 1992, ch. 35, U.N. Doc. A/Conf.151/26 (Vol. D III) (hereinafter Agenda 21).
[3] *Id.* ¶ 35.7(c).
[4] *Id.* ¶ 8.14.
[5] Report of the fist session of the Criteria Expert Group (CEG1) for POPs, Bangkok, (Oct. 26-30, 1998), at http://www.chem.unep.ch/sc/documents/meetings/ceg1/CEG1-3.htm (last visited on Jan. 28, 2002).

Chapter 2

b) Rotterdam Convention on the Prior Informed Consent Procedure for Certain Hazardous Chemicals and Pesticides in International Trade;

c) Protocol to the Convention on Long-range Transboundary Air Pollution on Persistent Organic Pollutants adopted under the auspices of the UN Economic Commission for Europe (ECE); and

d) Stockholm Convention on Persistent Organic Pollutants.

This section introduces the chemical structure and nomenclature of the twelve most persistent organic pollutants and their toxicity.

A. Pesticides

1. Aldrin

Chemical Structure

CAS IUPA[6] chemical name: 1,2,3,4,10,10-Hexachloro-1, 4,4a, 5,8,8a-hexahydro-1, 4:5,8-dimethanonaphthalene.

Synonyms and Trade Names (partial list): Aldrec, Aldrex, Aldrex 30, Aldrite, Aldrosol, Altox, Compound 118, Drinox, Octalene, Seedrin.

CAS No.: 309-00-2; molecular formula: C12H8Cl6; formula weight: 364.92

Physical Appearance: White, odorless crystals when pure; technical grades are tan to dark brown with a mild chemical odor.

Properties: Melting point: 104° C (pure), 49-60° C (technical); boiling point: 145° C at 2 mm Hg; KH: 4.96 x 10-4 atm m³/mol at 25° C; $\log_{KOC}$: 2.61, 4.69; $\log K_{OW}$: 5.17-7.4; solubility in water: 17-180 µg/L at 25° C; vapor pressure: 2.31 x 10-5 mm Hg at 20° C.

---

6 IUPA stands for a systematic set of rules adopted by Internal Union of Pure and Applied Chemists in 1960.

Aldrin is an insecticide used to control soil insects such as termites, corn rootworm, wireworms, rice water weevil, and grasshoppers. From 1950 to 1970 it was widely used as soil insecticide to control rootworms, beetles, and has been effective to protect wooden structures from termites.[7]

Aldrin is readily converted to dieldrin in the environment, and the two insecticides are closely related chemically. It binds strongly to soil particles and slowly evaporate to the air, so the most likely route this toxic substance enters the human body is by eating food grown in treated soil, by eating products from fish, poultry, or beef previously exposed to these insecticides, or by drinking water or milk containing residue.[8]

Due to its persistent nature and hydrophobicity, i.e., easily dissolved in fat, aldrin is known to bioconcentrate, mainly as its conversion products.

Toxicity: Aldrin is toxic to humans and mainly affects the central nervous system; the lethal dose of aldrin for an adult man has been estimated to be about 5g, equivalent to 83-mg/kg-body weight.[9] Signs and symptoms of aldrin intoxication may include headache, dizziness, nausea, general malaise, and vomiting, followed by muscle twitchings, myoclonic jerks, and convulsions. Occupational exposure to aldrin, in conjunction with dieldrin and endrin, was associated with a significant increase in liver and biliary cancer, although the study did have some limitations, including a lack of quantitative exposure information.[10] There is limited information that cyclodienes, such as aldrin, may affect immune responses.

The acute oral LD_{50} for aldrin in laboratory animals is in the range of 33-mg/kg body weights for guinea pigs to 320-mg/kg body weights for hamsters.[11] Reproductive effects in rats were observed when pregnant females were dosed with 1.0 mg/kg aldrin subcutaneously.[12] Offspring experienced a decrease in the median effective time for incisor teeth eruption and increase in the median effective time for testes descent. There is, as of yet, no evidence of a teratogenic potential for aldrin. The International Agency for Research on Cancer (IARC) has concluded that there is inadequate evidence for the carcinogenity of aldrin in humans, and there is only limited evidence in experimental animals.[13] Aldrin is, therefore, not classified as to its carcinogenity in humans (IARC, Group 3).[14]

---

7   Report, Agency for Toxic Substances and Disease Registry (ATSDR), *Toxicological Profile for Aldrin & Dieldrin*, (United States Public Health Service, Atlanta, GA) May 1989-1993.
8   See id.
9   See id.
10 See id.
11 J. HARTE, TOXICS A TO Z: A GUIDE TO EVERYDAY POLLUTION HAZARDS 207-209 (University of California Press 1991).
12 See id.
13 R. Fahring, *Mutagenic Studies on Pesticides,* in CHEMICAL CARCINOGESIS ASSAYS 10 (International Agency for Research on Cancer, Scientific Publication, 1974).
14 The International Agency for Research on Cancer (IARC) is part of the World Health Organization and its mission is to coordinate and conduct research on the causes of human cancer, the mechanisms of carcinogenesis, and to develop scientific strategies for cancer control. Its headquarters are in France. Web site: http://www.iarc.fr.

Aldrin has low phytotoxicity, which is its induced oxidation by light energy, with plants affected only by extremely high application rates.[15] The toxicity of aldrin to aquatic organisms is quite variable, with aquatic insects being the most sensitive group of invertebrates. The 96-h LD_{50} values range from 1-200 $\mu g/L$ for insects, and from 2.2-53 $\mu g/L$ for fish. Long term and bioconcentration studies are performed primarily using dieldrin, the primary conversion product of aldrin. In a model ecosystem study, only 0.5% of the original radioactive aldrin was stored as aldrin in the mosquito fish (*Gambusia affinis*), the organism at the top of the model food chain, the balance of which was converted to dieldrin.[16]

The acute toxicity of aldrin to avian species varies in the range of 6.6 mg/kg for bobwhite quail to 520 mg/kg for mallard ducks.[17] Aldrin-treated rice is thought to have been the cause of deaths for waterfowl, shorebirds and passerines along the Texas Gulf Coast, both by direct poisoning by ingestion of aldrin-treated rice and indirectly by consuming organisms contaminated with aldrin. Residues of aldrin were detected in all samples of bird casualties, eggs, scavengers, predators, fish, frogs, invertebrates and soil.[18]

Aldrin is banned in many countries, including Bulgaria, Ecuador, Finland, Hungary, Israel, Singapore, Switzerland and Turkey. Its use is severely restricted in many countries, including Argentina, Austria, Canada, Chile, Japan, New Zealand, the Philippines, and Venezuela. In the European Union aldrin is a prohibited pesticide and is subject to Prior Informed Consent for FAO and WHO countries.[19] In the United States the use of aldrin was suspended by the Environmental Protection Agency in 1974 on the basis of cancer risk.[20] However, the use of aldrin as subterranean termiticide continued after 1974. Its importation altogether ceased in 1985 and its registration was cancelled two years later.[21] Its use was completely banned in the United States in 1985.

Information above mentioned excerpted from:

> 1) Toxicological Profile for Aldrin & Dieldrin. Atlanta, GA.
> Agency for Toxic Substances and Disease Registry (ATSDR) May 1989 and 1993—United States Public Health Service.

> 2) Dictionary of Environmentally Important Chemicals
> D.C. Ayres and D.G. Hellier
> Queen Mary and Westfield College, University of London, UK
> Fitzroy Dearborn Publishers, London. 1998.

---

15   *See* Harte, *supra* note 76.
16   *See id.*
17   *See id.*
18   *See id.*
19   *See id.*
20   *See id.*
21   *See id.*

3) Assessment Report prepared for the IPCS on: DDT, Aldrin, Dieldrin, Endrin, Chlordane, Heptachlor, Hexachlorobenzene, Mirex, Toxaphene, Polychlorinated Biphenyls, Dioxins, and Furans.
Prepared by: L. Ritter, K.R. Solomon, and J. Forget.
Canadian Network of Toxicology Centers—December 1995.

4) United Nations Environment Programme—UNEP Chemicals: Official Web Site: http://www.chem.unep.ch/pops/indxhtms/asses6.html.

2. Dieldrin

Chemical Structure

CAS Chemical Name: 3,4,5,6,9,9-Hexachloro-1a,2,2a,3,6,6a,7,7a-octahydro-2,7:3,6-dimetanonapth[2,3-b]oxirene.

Synonyms and Trade Names (partial list): Alvit, Dieldrite, Dieldrix, Illoxol, Panoram D-31, Quintox.

CAS No.: 60-57-1; molecular formula: C12H8Cl6O; formula weight: 380.91.

Physical Appearance: A stereoisomer of endrin, dieldrin may be present as white crystals or pale tan flakes, odorless to mild chemical odor.

Properties: Melting point: 175°-176° C; boiling point: decomposes; K_h: 5.8 x 10-5 atm·m³/mol at 250 C; log K_{OC}: 4.08-4.55; log K_{OW}: 3.692-6.2; solubility in water: 140 μg/L at 20° C; vapor pressure: 1.78 x 10-7 mm Hg at 20° C.

Dieldrin is a synthetic organochlorine insecticide with a chemical structure similar to aldrin. Dieldrin is easily broken down from aldrin in soil, water and living organisms. Since 1950 it has been used for agricultural control of soil insects, other insect vectors of disease, and veterinary agents.[22] It has also been used in timber treatment and is effective against disease vectors such as locusts and the tsetse fly. It is a human carcinogen and extremely persistent in the environment. Studies in animals have demonstrated that dieldrin causes liver damage, central nervous

---

22   D. C. AYRES & D.G. HELLIER, DICTIONARY OF ENVIRONMENTALLY IMPORTANT CHEMICALS 129-130 (1998).

Chapter 2

effects and suppression of the immune system, as well as acting as a disruptor of the endocrine system, possibly in the form of an estrogen disruptor.[23] It has been banned in the U.S. since 1975 and has also been banned in many other parts of the world.[24]

Information above mentioned excerpted from:

1) Toxicological Profile for Aldrin & Dieldrin. Atlanta, GA.
Agency for Toxic Substances and Disease Registry (ATSDR) May 1989 and 1993
United States Public Health Service.

2) Dictionary of Environmentally Important Chemicals
D.C. Ayres and D.G. Hellier
Queen Mary and Westfield College, University of London, UK
Fitzroy Dearborn Publishers, London. 1998.

3) Assessment Report prepared for the IPCS on: DDT, Aldrin, Dieldrin, Endrin, Chlordane, Heptachlor, Hexachlorobenzene, Mirex, Toxaphene, Polychlorinated Biphenyls, Dioxins, and Furans.
Prepared by: L. Ritter, K.R. Solomon, and J. Forget.
Canadian Network of Toxicology Centers—December 1995.

4) United Nations Environmental Program—Chemicals: Official Web Site:
http://www.chem.unep.ch/pops/indxhtms/asses6.html.

3. Endrin

Chemical Structure

CAS Chemical Name: 3,4,5,6,9,9, -Hexachloro-1a, 2,2a, 3,6,6a, 7,7a-octahydro-2, and 7:3,6-dimethanonaphth [2,3-b] oxirene.

Synonyms and Trade Names (partial list): Compound 269, Endrex, Hexadrin, Isodrin Epoxide, Mendrin, Nendrin.

---

23 *See id.*
24 *See id. supra*, FN 11.

CAS No.: 72-20-8; molecular formula: C12H8Cl6O; formula weight: 380.92.

Physical Appearance: White, odorless, crystalline solid when pure; light tan color with faint chemical odor for technical grade.

Properties: Melting point: 200° C; boiling point: 245° C (decomposes); KH: 5.0 x 10-7 atm·m³/molecular; log K_{ow}: 3.209-5.339; solubility on water: 220-260 µg/L at 25° C; vapor pressure: 7 x 10-7 mm Hg at 25° C. Insecticide. Toxicity class I.

Endrin is a foliar insecticide used mainly on fields of crops such as cotton and grains. It has also been used as a rodenticide to control mice and voles, as a chemical to discourage birds from nesting on buildings, and as an insecticide. It is highly toxic to fish and fatal to humans and it is also persistent, having an estimated half-life in soil for more than 14 years.[25] This compound was introduced in the market in 1951 and its use has been banned in developed countries; in certain developing nations such as Brazil its use is severely restricted.[26] The United States banned its use in 1980.[27] Human exposure takes place mainly through the consumption of food or water contaminated with residues. Occupational contamination is also a reality. In human health effects include endocrine disruption, liver damage, and nervous system disorders.[28]

Cotton demands the largest proportion of pesticide use in the world's agricultural activities. In 1968, 86% of endrin was used in the cotton fields in the United States.[29] During its use for 10 years in the United States, endrin caused the death of enormous populations of fish and poisoned cattle that walked through sprayed orchards.[30]

Information above mentioned excerpted from:

1) Dictionary of Environmentally Important Chemicals
D.C. Ayres and D.G. Hellier
Queen Mary and Westfield College, University of London, UK
Fitzroy Dearborn Publishers, London. 1998.

2) Our Children's Toxic Legacy: How Science and Law Fail to Protect Us From Pesticides. John Wargo. Second Edition. Yale University Press. New Haven. 1998.

3) Silent Spring. Rachel Carson. Houghton Mifflin Co. New York. 1962.

---

25 Linda Ritter, et al. *Assessment Report on DDT, Aldrin, Dieldrin, Endrin, Chlordane, Heptachlor, Hexachlorobenzene, Mirex, Toxaphene, Polychlorinated Biphenyls, Dioxins, and Furans*, 1st Session of the INC, UNEP/POPS/INC.1/INF10 (1998), *available at* http://irptc.unep.ch/pops/POPs_/Inc_1/RITTER-En.html. (last visited on April 5th, 2002).
26 Telephone interview with Júlio Sérgio de Britto, Brazilian Ministry of Agriculture (Oct. 12th, 2001).
27 *See id., supra* FN 95.
28 *See id.*
29 J. Wargo, Our Children's Toxic Legacy: How Science and Law Fail to Protect Us from Pesticides 51 (2d ed. 1998).
30 R. Carson, Silent Spring 25-27 (1962).

4) Assessment Report prepared for the IPCS on: DDT, Aldrin, Dieldrin, Endrin, Chlordane, Heptachlor, Hexachlorobenzene, Mirex, Toxaphene, Polychlorinated Biphenyls, Dioxins, and Furans.
Prepared by: L. Ritter, K.R. Solomon, J. Forget.
Canadian Network of Toxicology Centers—December 1995.

5) United Nations Environment Programme—UNEP Chemicals: Official Web Site: http://www.chem.unep.ch/pops/indxhtms/asses6.tml

4. Chlordane

Chemical Structure

Cis   Trans

CAS Chemical Name: 1,2,4,5,6,7,8,8-octachloro-2,3,3a,4,7,7a-hexahydro-4,7-methano-1H-indene.

Trade names: (partial list): Aspon, Belt, Chloriandin, Chlorkil, Chlordane, Corodan, Cortilan-neu, Dowchlor, HCS 3260, Kypchlor, M140, Niran, Octachlor, Octaterr, Ortho-Klor, Synklor, Tat chlor 4, Topichlor, Toxichlor, Veliscol-1068.

CAS No.: 57-74-9; molecular formula: C10H6Cl8; formula weight: 409.78.

Physical Appearance: colorless to yellowish-brown viscous liquid with an aromatic, pungent odor similar to chlorine.

Properties: Melting point: ° C; boiling point: 165° C at 2 mm Hg; KH: 4.8 x 10-5 atm m³/mol at 25° C; log K_{OC}: 4.58-5.57; log K_{OW}: 6.00; solubility in water: 56 ppb at 25° C; vapor pressure: 10-6 mm Hg at 20° C.

Similar to many of the chlorinated synthetic pesticides known as persistent organic pollutants, chlordane was used for many years as an insecticide. It is known as a culprit for its toxic effects and its capacity to persist and bioaccumulate in the environment.[31] This broad-spectrum organochlorine compound was introduced in 1945 for controlling termites and soil-borne insects whose larvae feed on the roots of plants and consequently cause great loss for farmers.[32] It was also used as a pesticide on corn, citrus, small grains, cotton, sugar beets, potatoes, sugarcane, and other crops.

Its biochemical properties are that it is highly insoluble in water, semi-volatile and, hence, bioconcentrates in the fat of living species, whether animal or humans. It is harmful on skin contact or if swallowed. There is not much controversy over the adverse effects or that chlordane is a carcinogenic as determined by the International Agency for Research on Cancer in animals.[33] The United States Environmental Protection Agency has categorized it under "probable human carcinogen" status.[34] Its cumulative toxicity also causes kidney and liver damages. Most scientists concur that it causes leukemia, aplastic anemia, convulsions, miscarriages, and birth defects in humans.

The United States Environmental Protection Agency banned all agricultural uses of chlordane in 1978,[35] allowing its use only for termite control, assuming that when used to exterminate termites, this substance would not come into contact with humans. For years alarming quantities of chlordane were being detected in the air. In 1983, the New York State Health Department confirmed high levels of chlordane in homes sampled.[36] A legal battle was fought between Velsicol Chemical Corporation and EPA, where Velsicol continued to reject the fact that there are dangers associated with chlordane.[37] Even though it was banned in 1978 for agricultural use, substantial amounts of chlordane are still being found on potatoes and beef. Dr Arnold Lehman, chief pharmacologist at the U.S. Food and Drug Administration, defined chlordane as: "one of the most toxic of all insecticides—anyone handling it could be poisoned."[38] Based on sample evidence brought by scientists, EPA finally made an agreement in 1987[39] with Velsicol Chemical Corporation, the proprietor and sole manufacturer of chlordane, to halt sales by April 15th, 1988.[40]

---

31 See Ritter, *supra* note 95.
32 See id.
33 See id.
34 Report, U.S. EPA, *Carcinogenicity Assessment of Chlordane and Heptachlor/Heptachlor Epoxide* (Office of Health and Environmental Assessment, Carcinogen Assessment Group 1986).
35 Velsicol Chemical Co., et al., *Consolidated heptachlor/chlordane Cancellation Proceedings*, 43 Fed. Reg. 12,372-12,379 (Apr. 12th, 1978).
36 THEO COLBORN ET AL., OUR STOLEN FUTURE 102 (1997).
37 61. U.S. v. Velsicol Chemical Corp., 701 F.Supp. 140, 29 ERC 1654 W.D.Tenn. Sep 15, 1987.
38 Arnold Lehman, *Dominant Lethal Studies with Technical Chlordane, HCS-3260, and Heptachlor: heptachlor epoxide, in* 2 J. TOXICOL. ENVTL. HEALTH 547-555 (1977).
39 Chlordane and Heptachlor Termicides, Cancellation Order, 52 Fed. Reg. 42,145-42,149 (1987).
40 Chlordane/Heptachlor Termicides, Notification of Cancellation and Amendment of Existing Stocks Determination, 53 Fed. Reg. 11,798-11,805 (1988).

Under the agreement, Velsicol Chemical Company in Memphis, Tennessee is prohibited from selling chlordane in the United States. The company still manufactures this chemical compound. In 1991, Velsicol produced and exported over 1.1 million pounds of chlordane.[41] Velsicol was the only American manufacturer of chlordane at the time that EPA cancelled its registration for commercial production, delivery, sale, and use in the U.S.[42] It is prohibited in the European Union and FAO/WHO countries.

Information above mentioned excerpted from:

1) Dictionary of Environmentally Important Chemicals
D.C. Ayres and D.G. Hellier
Queen Mary and Westfield College, University of London, UK
Fitzroy Dearborn Publishers, London. 1998.

2) Assessment Report prepared for the IPCS on: DDT, Aldrin, Dieldrin, Endrin, Chlordane, Heptachlor, Hexachlorobenzene, Mirex, Toxaphene, Polychlorinated Biphenyls, Dioxins, and Furans.
Prepared by: L. Ritter, K.R. Solomon, J. Forget.
Canadian Network of Toxicology Centers—December 1995.

3) United Nations Environment Programme—UNEP Chemicals: Official Web Site: http://www.chem.unep.ch/pops/indxhtms/asses6.html.

5. Dichlorodiphenyltrichloroethane (DDT)

Chemical Structure

CAS Chemical Name: 1,1'-(2,2,2-Trichloroethylidene)bis(4-chlorobenzene)

Synonyms and Trade Names (partial list): Agritan, Anofex, Arkotine, Azotox, Bosan Supra, Bovidermol, Chlorophenothan, Chloropenothane, Clorophenotoxum, Citox, Clofenotane, Dedelo, Deoval, Detox, Detoxan, Dibovan, Dichlorodiphenyltrichloroethane, Dicophane, Didigam, Didimac, Dodat, Dykol, Estonate,

---

41 M. Zimmerman, Science, Nonscience and Nonsense Approaching Environmental Literacy 58-61 (1997).
42 Chlordane/Heptachlor Termicides, Notification of Cancellation and Amendment of Existing Stocks Determination, 53 Fed. Reg. 11,798-11,805 (12th July, 1988).

Genitox, Gesafid, Gesapon, Gesarex, Gesarol, Guesapon, Gyron, Havero-extra, Ivotan, Ixodex, Kopsol, Mutoxin, Neocid, Parachlorocidum, Pentachlorin, Pentech, PPzeidan, Rudseam, Santobane, Zeidane, Zerdane.

CAS No.: 50-29-3; molecular formula: C14H9Cl5; formula weight: 354.49.

Physical Appearance: Odorless to slightly fragrant colorless crystals or white powder. It may also be found as aerosol, emulsifiable concentrate, and wettable powder.

Properties: Melting point: 108.5° C; boiling point: 185° C at 0.05 mm Hg (decomposes); KH: 1.29 x 10-5 atm·m³/mol at 23° C; log K_{OC}: 5.146-6.26; log K_{OW}: 4.89-6.914; solubility in water: 1.2-5.5 μg/L at 25° C.

Othmar Zeidler first synthesized this chemical substance, **dichlorodiphenyltrichloroethane**, commonly known as DDT, in 1874. At the time Zeidler was a student at the University of Strasbourg. It was a Swiss Chemist, Paul Müller, who was working for the German chemical company Geigy AG, who rediscovered its properties as an insecticide in 1939 while searching for a contact poison against clothes moths and carpet beetles.[43] DDT is a contact poison that acts on the nervous system of insects, causing over stimulation of neurons and rapid death. The effectiveness of DDT against a variety of household and crop insect pests was quickly demonstrated, earning Paul Müller a Nobel Prize in 1948 for his discoveries.[44]

Insect-borne diseases were killing more American soldiers than the enemy during the war around 1943.[45] Before the end of World War II, DDT was available to the Allies and was first used medically in the suppression of a typhus epidemic in Napes, Italy, during the winter of 1943–1944, when it was applied directly to humans to control lice.[46]

At that time, DDT was less toxic to humans than the crude lead-arsenate insecticides then in use. One of the appreciated properties of DDT is that it could kill hundreds of pests and stay lethal long after it was sprayed.

DDT's efficacy and low-production costs made it the most popular insecticide worldwide from 1946 to 1972, with an estimated annual world production in this period of 2.8 million tons.[47] In the United States alone, American chemical industries were producing nearly 2 million pounds of DDT per month in 1944 to

---

[43] See Wargo, *supra* note 99.
[44] *DDT: Nobel Prize in Medicine for Dr. Paul Mueller* 1948 Senior Scholastic 53 (Nov. 10th).
[45] THE SCIENCE OF LIFE: CONTRIBUTIONS OF BIOLOGY TO HUMAN WELFARE 243-294 (K.D. Fisher & A.U. Nixon, eds., 1972).
[46] G.T. Brooks, *Chlorinated Insecticides. Technology and Application*, 1 CRC 12-13 (1974).
[47] Meeting Background Report, International Experts Meeting on Persistent Organic Pollutants (1995) (Annex II: Profile Substances Draft, Vancouver, B.C.).

fulfill the needs of the military.[48] DDT was mandated for military use only until the end of the war in 1945.[49] But soon thereafter, DDT's potential effectiveness for other applications was discovered. Its broad scale production and use in agriculture saved millions of lives which otherwise might have been lost to starvation.

One of the most popular uses of DDT since its discovery has been to deal with insect-borne diseases such as malaria. Malaria is caused by a tiny little parasite hosted by female *Antopheles* mosquitoes. It has been feared around the world as being the most deadly disease in modern history. Malaria has not only killed hundreds of thousands of humans but also has infected and debilitated hundreds of millions each year.[50] To put it into perspective, World War II killed some 15 million people.[51] In 1930, almost 50 million people died worldwide of malaria, which made the discovery of DDT a miracle for the time.[52] The great majority of malaria cases are concentrated in developing countries where no integrated national malaria management program has ever been implemented.[53] In 1991, the WHO estimated that almost ten per cent of the world's population lives in areas where malaria continues to rage.[54] The WHO continues to approve the use of DDT for the control of malaria, provided some pre-established criteria are met, due to the emergent need, low production cost, and efficacy of the pesticide regardless of its effects on human health.[55]

DDT has been and continues to be banned in most developed countries and elsewhere where malaria is not a threat. In developing countries such as Brazil, DDT is banned except for vector control of malaria, meaning that it may be used in all regions where the disease poses a threat to public health.[56] In 1949, scientific data proved that DDT was a definite health hazard to humans.[57] The United States Department of Agriculture decided to prohibit the use of DDT in cattle because it had been demonstrated that significant amount of DDT was stored in fat and milk for nearly six months after a cow's last expo-

---

[48] RACHEL CARSON, SILENT SPRING, (Houghton Mifflin, 1962).
[49] E.F. Knipling, *DDT Insecticides Developed for Use by the Armed Forces* 38 J. ECON. ENTERPRISE 201 (1945).
[50] W.J. HAYES, HANDBOOK OF PESTICIDE TOXICITY 11 (1991).
[51] D.J. SINGER & M. SMALL, THE WAGES OF WAR 1816-1965: A STATISTICAL HANDBOOK (1986).
[52] *See id.*
[53] World Health Organization, *World Malaria Situation in 1990*, 6 WHO BULLETIN 801-07 (1992).
[54] *See id.*
[55] WORLD HEALTH ORGANIZATION, ENVIRONMENTAL HEALTH CRITERIA 83, DDT AND ITS DERIVATES, ENVIRONMENTAL ASPECTS, (1989).
[56] Telephone interview with Dr. Júlio Sérgio de Britto, Brazilian Ministry of Agriculture (Oct. 12th, 2001).
[57] SANDRA STEINGRABER, LIVING DOWNSTREAM: A SCIENTIST'S PERSONAL INVESTIGATION OF CANCER AND THE ENVIRONMENT 9-10 (1997).

sure to DDT.[58] A huge battle raged for years between chemical companies and governments around the world, one side claiming that DDT was safe for use and the other claiming that DDT was extremely hazardous to the environment and to human health.[59]

In the summer of 1962, Rachel Carson of Massachusetts, an American wildlife biologist, published three articles in the New Yorker magazine warning the population of the toxic effects of pesticides. Later that year, Rachel Carson published a book entitled *Silent Spring*, in which she stated that the risk of pesticides to our health and to the environment was comparable to the risks of a nuclear war.[60] In her book, Carson presented technical arguments carefully supported by evidence and backed up by data collected from scientists. She tried to make it clear to the public that humans are connected to their surrounding environment through their food supply, and she explained the facts of persistence and accumulation of DDT.[61] The United States banned the domestic usage of DDT on June 14th 1972, by EPA administrator William D. Ruckelshaus.[62]

Information above mentioned excerpted from:

1) Dictionary of Environmentally Important Chemicals
D.C. Ayres and D.G. Hellier
Queen Mary and Westfield College, University of London, UK
Fitzroy Dearborn Publishers, London. 1998.

2) Brooks, GT. 1974. Chlorinated Insecticides. Technology and Application. Cleveland, Ohio. CRC: Vol. 1, pages 12-13.

3) Assessment Report prepared for the IPCS on: DDT, Aldrin, Dieldrin, Endrin, Chlordane, Heptachlor, Hexachlorobenzene, Mirex, Toxaphene, Polychlorinated Biphenyls, Dioxins, and Furans. Prepared by: L. Ritter, K.R. Solomon, and J. Forget.
Canadian Network of Toxicology Centers—December 1995.

4) Wargo, J. 1998. Our Children's Toxic Legacy: How Science and Law Fail to Protect Us from Pesticides. Second Edition. Yale University Press. New Haven.

United Nations Environment Programme—UNEP Chemicals: Official Web Site: http://www.chem.unep.ch/pops/indxhtms/asses6.html.

5) Carson, R. 1962. Silent Spring. Houghton Mifflin CO. NY.

---

58 J.B. Shepherd, et al., *The Effect of Feeding Alfalfa Hay Containing DDT Residue on the DDT Content of Cow's Milk* 23 J. OF DAIRY SCI. 549-55 (1949).
59 See id., supra note 121, at 114-117.
60 S. Udall, *The Legacy of Rachel Carson* 1964 SATURDAY REVIEW 23.
61 R. CARSON, SILENT SPRING 136-37 (1962).
62 Press Release, EPA, *DDT ban takes effect* (Dec. 31st, 1972) at http://www.epa.gov/history/topics/ddt/01.htm.

6. Heptachlor

Chemical Structure

CAS Chemical Name: 1,4,5,6,7,8,8-Heptachloro-3a, 4,7,7a-tetrahydro-4, 7-methanol-1*H*-indene.

Synonyms and Trade Names (partial list): Aahepta, Agroceres, Baskalor, Drinox, Drinox H-34, Heptachlorane, Heptagran, Heptagranox, Heptamak, Heptamul, Heptasol, Heptox, Soleptax, Rhodiachlor, Veliscol 104, Veliscol heptachlor.

CAS No.: 76-44-8; molecular formula: C10H5Cl7; formula weight: 373.32.

Physical Appearance: White to light tan, waxy solid or crystals with a camphor-like odor.

Properties: Melting point: 95°-96°C (pure), 46°-74° C (technical); boiling point: 135°-145° C at 1-1.5 mm Hg, decomposes at 760 mm Hg; KH; 2.3 x 10-3 atm·mm³/mol; log K_{OC}: 4.38; log K_{OW}; 4.40-5.5; solubility in water: 180 ppb at 25° C; vapor pressure: 3 x 10-4 mm Hg at 20° C.

Heptachlor is a non-systemic stomach and contact insecticide first introduced by Velsicol Chemical for the treatment of ants and termites in soil and for household insects. It is highly insoluble in water but is soluble in organic solvents. It binds readily to aquatic sediments and bioconcentrates in the fat of living organisms. It travels through the food chain on particles of fat or vapor that are transported by the wind to distant places. For instance, scientific investigations on Signy Island, in Antarctica, uncovered high concentration residues of heptachlor in penguins and skuas—a scavenger bird that migrates to the tropics—as well as on krill, which is the most common food for the blue whale.[63] Heptachlor was banned in the United States in 1983.[64]

---

63 J.H. Tatton, *Organochlorine Pesticides in Antarctica* 215 NATURE 346-48 (1967).
64 S. STEINGRABER, LIVING DOWNSTREAM: A SCIENTIST'S PERSONAL INVESTIGATION OF CANCER AND THE ENVIRONMENT 9-10 (1997).

Information above mentioned excerpted from:

1) Dictionary of Environmentally Important Chemicals
D.C. Ayres and D.G. Hellier
Queen Mary and Westfield College, University of London, UK
Fitzroy Dearborn Publishers, London. 1998.

2) Assessment Report prepared for the IPCS on: DDT, Aldrin, Dieldrin, Endrin, Chlordane, Heptachlor, Hexachlorobenzene, Mirex, Toxaphene, Polychlorinated Biphenyls, Dioxins, and Furans.
Prepared by: L. Ritter, K.R. Solomon, J. Forget.
Canadian Network of Toxicology Centers—December 1995.

3) United Nations Environment Programme—UNEP Chemicals:
Official Website: http://www.chem.unep.ch/pops/indxhtms/asses6.html.

7. Mirex

Chemical Structure

CAS chemical name: 1,1a, 2,2,3,3a, 4,5,5a, 5b, 6-dodecachloroacta-hydro-1, 3,4-metheno-1H-cyclobuta [cd] pentalene.

Synonyms and Trade Names (partial list): Dechlorane, Ferriamicide, GC 1283.

CAS N0: 2385-85-5; molecular formula: C10Cl12; formula weight: 545.5.

Appearance: White crystalline, odourless solid.

Properties: Melting point: 485° C; vapor pressure: 3×10^{-7} mm Hg at 25° C.

Mirex was first synthesized on a laboratory scale in 1946. Its industrial production began in the United States in 1959 by Allied Chemical Company under the name GC-1283.[65] This snow-white, odorless, crystalline solid is a synthetic insec-

---

65 *See* Ayres, *supra* note 92, at 208-209.

ticide[66] widely used in the United States from 1960 to 1970 to control fire ants, especially in the Southeast. Mirex has also been used to combat leaf cutter ants in South America, harvester termites in South Africa, Western harvester ants in the U.S., and pineapple mealybug in Hawaii.[67]

Mirex was employed as a flame retardant additive under the trade name Dechlorane[68] in plastics, rubber, paint, paper, and electrical goods from 1959 to 1972.[69] It was produced and sold in the United States by Hooker Chemical Company, Niagara Falls, New York from 1957 to 1976.[70] It is no longer produced commercially in the United States and EPA cancelled the registration of pesticides containing mirex under FIFRA, with specified termination of uses of existing stock.

Mirex is very persistent and of extreme biomagnification capability. Due to its low water solubility, high lipid solubility, high stability, and semi-volatility, it can be transported to distant locations and studies have demonstrated residues in Arctic freshwater and terrestrial organisms.

Information above mentioned excerpted from:

1) Dictionary of Environmentally Important Chemicals
D.C. Ayres and D.G. Hellier
Queen Mary and Westfield College, University of London, UK
Fitzroy Dearborn Publishers, London. 1998.

2) Budavari, S.; O'Neil, M. J.; Smith, A.; et al. Eds. Merck. 1989. The Merck Index: An Encyclopedia of Chemicals, Drugs, and Biologicals. 11th Edition, Rahway, NJ. Merck and Co, Inc. 321, 977.

3) Assessment Report prepared for the IPCS on: DDT, Aldrin, Dieldrin, Endrin, Chlordane, Heptachlor, Hexachlorobenzene, Mirex, Toxaphene, Polychlorinated Biphenyls, Dioxins, and Furans. Prepared by: L. Ritter, K.R. Solomon, J. Forget.
Canadian Network of Toxicology Centers—December 1995.

4) United Nations Environment Programme—UNEP Chemicals: Official Website: http://www.chem.unep.ch/pops/indxhtms/asses6.html.

---

66 World Health Organization, *Monographs on the Evaluation of Carcinogenic Risk of Chemicals to Humans* 20 International Agency for Research on Cancer 283-301 (1979).
67 D.A. Carlson, et al., *Mirex in the Environment: Its degradation to Kepone and Related Compounds* 194 SCIENCE 939-41 (1976).
68 Designation of Hazardous Substances, 40 C.F.R. 116 (1978).
69 THE MERCK INDEX: AN ENCYCLOPEDIA OF CHEMICALS, DRUGS, AND BIOLOGICALS 321, 977 (11th ed., S. Budavari, et al. eds., 1989).
70 T.W. Lewis, & J.C. Makarewicz, *Exchange Mirex Between Lake Ontario USA and its Tributaries* 14 J. OF GREAT LAKE RESEARCH 388-93 (1988).

8. Toxaphene

Chemical Structure

CAS Chemical Name: Toxaphene

Synonyms and Trade Names (parital list): Alltex, Alltox, Attac 4-2, Attac 4-4, Attac 6, Attac 6-3, Attac 8, Camphechlor, Camphochlor, Camphoclor, Chemphene M5055, chlorinated camphene, Chloro-camphene, Clor chem T-590, Compound 3956, Huilex, Kamfochlor, Melipax, Motox, Octachlorocamphene, Penphene, Phenacide, Phenatox, Phenphane, Polychlorocamphene, Strobane-T, Strobane T-90, Texadust, Toxakil, Toxon 63, Toxyphen, Vertac 90%.

CAS No.: 8001-35-2; molecular formula: C10H10Cl8; formula weight: 413.82.

Physical Appearance: Yellow, waxy solid with a chlorine/terpene-like odor.

Properties: Melting point: 65°-90° C; boiling point: >120° C (decomposes); KH: 6.3 x 10-2 atm·m³/molecular; log K_{OC}: 3.18 (calculated); log K_{OW}: 3.23-5.50; solubility in water: 550 µg/L at 20° C; vapor pressure: 0.2-0.4 mm Hg at 25° C.

This non-systemic contact insecticide named toxaphene was first used in 1948 for the control of grasshoppers.[71] It was probably the world's most widely used pesticide in the seventies to control insect pests on cotton, cereal grains, fruits, nuts, and vegetables. In addition, toxaphene was also applied to eliminate certain fish species that were then considered by some to be undesirable.[72]

Another use of this chemical compound was to control ticks, flies, mange, and mites on livestock and poultry.[73] Toxaphene was the most heavily employed insecticide in the United States until 1982.[74] In 1990, the U.S. EPA banned all uses

---

71 *See* Ayres, *supra* note 92, at 305-306.
72 W.L. Lockhart, et al., *Presence and Implications of Chemical Contaminants in Fresh Water of the Canadian Arctic* 122 SCI. TOTAL ENV'T. 165-243 (1992).
73 E.F. Knipling & W.E. Westlake, *Insecticide Use in Livestock Production* 13 RESIDUE REV. 1-32 (1966).
74 47 Fed. Reg. 53,784-53,793 (1982).

of this chemical due to concrete, scientific evidence demonstrating its adverse effects to humans and animals.[75]

Toxaphene is, however, still produced and used in many developing countries, such as India, many countries in Latin America, Eastern Europe, and Africa.[76] A recent study concludes that the global toxaphene consumption is greater than 1.3 million tons for the period 1950 to 1993. [77]

Information above mentioned excerpted from:

> 1) Dictionary of Environmentally Important Chemicals
> D.C. Ayres and D.G. Hellier
> Queen Mary and Westfield College, University of London, UK
> Fitzroy Dearborn Publishers, London. 1998.
>
> 2) Assessment Report prepared for the IPCS on: DDT, Aldrin, Dieldrin, Endrin, Chlordane, Heptachlor, Hexachlorobenzene, Mirex, Toxaphene, Polychlorinated Biphenyls, Dioxins, and Furans.
> Prepared by: L. Ritter, K.R. Solomon, J. Forget.
> Canadian Network of Toxicology Centers - December 1995.
>
> 3) United Nations Environment Programme—UNEP Chemicals: Official Website: http://www.chem.unep.ch/pops/indxhtms/asses6.html.
>
> 4) Agency for Toxic Substances and Disease Registry of the U. S. Department of Health and Human Services Website:
> http://www.atsdr.cdc.gov/toxprofiles/tp94.pdf.

B. Industrial Chemicals

1. Polychlorinated Biphenyls—PCBs

Chemical Structure

Trade Names for different mixtures (partial list): Aroclor, Pyranol, Pyroclor, Phenochlor, Pyralene, Clophen, Elaol, Kanechlor, Santotherm, Fenchlor, Apirolio, Sovol.

CAS No.: 1336-36-3.

---

75 55 Fed. Reg. 31,164-31,174 (1990).
76 T.F. Bidleman, et al., *Toxaphene and other Organochlorines in Arctic Ocean Fauna: Evidence for Atmospheric Delivery* 42 Arctic 307-13 (1989).
77 E.C. Voldner & Y.C. Li, *Global Usage of Toxaphene* 27 Chemosphere 2073-78 (1993).

This chemical compound comprises a group of 209 compounds synthesized by two aromatic rings of cyclic carbon atoms referred to as biphenyl welded directly together, around which there are varying numbers of chlorine atoms. That is why there are 209 possible combinations of different forms of PCBs. Its main characteristics are low flammability, low electrical conductivity (which makes PCBs perfect for use in insulators), high resistance to thermal breakdown and to other chemical agents, a high degree of chemical stability and, above all extremely low solubility in water. Therefore, it has been extremely successful as a viscous fluid used in electrical transformers, large capacitors, and small electronic parts as an insulator. Its use is, however, much broader since PCBs have also been used as heat exchange fluids, paint additives, in carbonless copy paper and in plastics.

PCBs were first manufactured commercially in 1929 in the United States, the United Kingdom, China, Germany, and Japan, and were exported throughout the world. In the United States, PCBs production was halted in 1977, at the time when it was being used predominantly as coolants and lubricants in electrical equipment.[78] The manufacture of PCBs was suspended due to evidence brought forward to Federal authorities demonstrating that PCBs not only accumulate but also persist in the environment, leading to toxic effects that threaten the health of humans and animals. During these 47 years of PCBs production and use in the United States, Monsanto Chemical Company manufactured approximately 99% of the PCBs at their production facility in Sauget, Illinois, and over 571,000 metric tons of PCBs were made available to the American market. PCB's were marketed worldwide under trade names such as Aroclor, Askarel, and Therminol.

Even though PCBs have been banned in most developed countries, they can, nevertheless, enter the environment from accidental spills, leaks during transportation of chemicals, fires in transformers containing PCBs, leaks from poorly maintained transformers, illegal or improper dumping of PCB waste into landfills or the burning of some municipal waste, illegal discharges and industrial incinerators. Once released into the environment, PCBs are not readily dispersed but will remain for very long periods of time.

Perhaps the two most serious accidents involving PCBs are the Yusho[79] incident in Japan in 1968, which exposed a large number of people through consumption of contaminated rice oil,[80] and the Yu-Cheng incident in Taiwan in 1979, through consumption of contaminated fish. On February 5th, 1968, a shipment of canned rice oil was thought to be the cause of contaminating 1,684 people who lived in the prefecture of Fukoaka, Japan. Samples of the cargo were taken for analysis, and PCBs in high concentration were found due to the use of

---

78 *See* Ritter, *supra* note 95, at 382-385.
79 Yusho in Japanese means literally "oil disease."
80 H. Tsukamoto, *The Chemical Studies on Detection of Toxic Compounds in the Rice Bran Oils used by the Patients of Yusho*, 60 FUKUOKA ACTA MED 496-512 (1969).

Kanechlor 400, a type of PCB produced by Kanegafuchi Chemical Industry Company.[81] In October of 1968, a study conducted by a group of scientists from the University of Kyushu, in Japan, showed the presence of PCBs not only in the oil but also in tissue samples from the patients examined, both live and deceased.[82] Children born up to 7 years after their mothers were exposed to PCBs in the Taiwan incident had deformed nails and natal teeth, intrauterine growth delay, and behavioral problems.

Human exposure to PCBs is normally through the consumption of contaminated food or water. Because of this accumulation and the widespread contamination of the food chain, virtually everyone has measurable levels of PCBs in their fatty tissues.[83] PCBs are toxic to aquatic organisms. In 1988, a team of veterinarians detected tumors in the body of four dead beluga whales in the Saint Lawrence estuary. These whales were struggling to reproduce, and cancer tumors were found in their intestines, breasts, bladders, ovaries, and stomachs. In all autopsied mammals, high levels of PCBs and other chlorinate pesticides were recorded.[84] A recent case involves PCBs in the Hudson River in New York, discharged over many years by General Electric Company. The United States Environmental Protection Agency in 2001 ordered General Electric to dredge the river to clean up PCBs in the sediment that have contaminated fish such as the striped bass in the river.[85]

Information above mentioned excerpted from:

1) Dictionary of Environmentally Important Chemicals
D.C. Ayres and D.G. Hellier
Queen Mary and Westfield College, University of London, UK
Fitzroy Dearborn Publishers, London. 1998.

2) Assessment Report prepared for the IPCS on: DDT, Aldrin, Dieldrin, Endrin, Chlordane, Heptachlor, Hexachlorobenzene, Mirex, Toxaphene, Polychlorinated Biphenyls, Dioxins, and Furans.
Prepared by: L. Ritter, K.R. Solomon, J. Forget.
Canadian Network of Toxicology Centers—December 1995.

3) United Nations Environment Programme—UNEP Chemicals: Official Website: http://www.chem.unep.ch/pops/indxhtms/asses6.html.

4) Agency for Toxic Substances and Disease Registry of the U. S. Department of Health and Human Services Website:
http://www.atsdr.cdc.gov/toxprofiles/tp17.html.

---

[81] M. Duratsume, *Epidemiological Studies on Yusho*, in PCB POISONING AND POLLUTION 9-23 (Kentaso Higushi ed., 1976).

[82] *See id.*

[83] W.J. Nicholson, & P.J. Landrigan, *Human Effects of Polychlorinated Biphenyls*, in DIOXINS AND HEALTH (A. Schecter ed. 1994).

[84] S. STEINGRABER, LIVING DOWNSTREAM: A SCIENTIST'S PERSONAL INVESTIGATION OF CANCER AND THE ENVIRONMENT 132-135 (1998).

[85] Telephone interview with Kevin Madonna, Attorney at Pace Environmental Law Clinic (Jan. 28th, 2002).

2. Hexachlorobenzene—HCB

Chemical Structure

CAS Chemical Name: hexachlorobenzene

Trade names: (partial list): Amaticin, Anticarie, Bunt-cure, Bunt-no-more, Co-op hexa, Granox, No bunt, Sanocide, Smut-go, Sniecotox

CAS No.: 118-74-1; molecular formula: C6Cl6; formula weight: 284.78;

Appearance: White monoclinic crystals or crystalline solid

Properties: Melting point: 227°-230° C; boiling point: 323°-326° C (sublimes); KH: 7.1 x 10-3 atm m³/mol at 20° C; log K_{OC}: 2.56-4.54; log K_{OW}: 3.03-6.42; Solubility in water: 40 µg/L at 20° C; vapor pressure: 1.089 x 10-5 mm Hg at 20° C.

Hexachlorobenzene is a chlorinated hydrocarbon industrial chemical that was manufactured by Lorentz in 1893.[86] In 1945, it was first introduced as a fungicide for seed treatment, especially of onions, wheat, and sorghum. It was widely used as a pesticide until 1965. It was also used to make fireworks, ammunition, and synthetic rubber. It is obtained industrially by the action of chlorine FeCl$_3$ on benzene at 150-200°C, or by recovery from waste tar from the production of tetrachloroethylene, which also contains heptachlorobutadiene.[87]

Hexachlorobenzene is among the most persistent environmental pollutants because of its chemical stability and resistance to degradation. It remains in the environment for a long time. Due to its high insolubility in water, most of it will remain in sediments on the bottom of lakes, rivers, or streams, posing serious threat to aquatic living creatures. Such particles may also be carried away in the air to long distances. This substance bioaccumaulates in the fat of living organisms as a result.

---

86 *See* Ayres, *supra* note 92, at 159-160.
87 *See id.*

Hexachlorobenzene is not currently manufactured as a commercial end product in the United States, where its production has been halted since the late seventies. However, hexachlorobenzene continues to be produced as a by-product or impurity in the manufacture of several chlorinated compounds and pesticides. Developing countries that still use hexachlorobenzene have contributed to its release in the environment, its improper storage and its disposal.

Information above mentioned excerpted from:

1) Dictionary of Environmentally Important Chemicals
D.C. Ayres and D.G. Hellier
Queen Mary and Westfield College, University of London, UK
Fitzroy Dearborn Publishers, London. 1998.

2) Toxicological Profile for Hexachlorobenzene
United States Department of Health and Human Services, Agency for Toxic Substances and Disease Registry
September 2000 by Syracuse Research Corporation
(contract # 205-1999-00024).

3) Assessment Report prepared for the IPCS on: DDT, Aldrin, Dieldrin, Endrin, Chlordane, Heptachlor, Hexachlorobenzene, Mirex, Toxaphene, Polychlorinated Biphenyls, Dioxins, and Furans.
Prepared by: L. Ritter, K.R. Solomon, J. Forget.
Canadian Network of Toxicology Centers—December 1995.

4) United Nations Environment Programme—UNEP Chemicals:
Official Website: http://www.chem.unep.ch/pops/indxhtms/asses6.html.

C. Unintended Produced By-products and Contaminants

1. Polychlorinated Dibenzo—p—Dioxins and Furans

Chemical Structure

2,3,7,8-TCDD   2,3,7,8-TCDF

Polychlorinated dibenzo-para-dioxins ("PCDD" or dioxin) and polychlorinated dibenzofurans ("PCDF" or furans) are two groups of planar tricyclic compounds that have very similar chemical structures and properties.[88] There are many who regard dioxin as a single compound, especially as it is remembered as the substance used to make "Agent Orange," which was widely used during the Vietnam War. Agent Orange was responsible for the death of many Americans and Vietnamese. As a matter of fact there are 75 different dioxins and 135 different furans that are distinguished by the position with a number of chlorine atoms attached to the two benzene rings. These different forms of dioxins are called isomers and they are similar in their distribution, toxic properties, and behavior.[89] Both are produced inadvertently during chemical manufacture and processing or by fires and incineration of municipal or industrial waste.

Their toxic effects depend both on the number of chlorine atoms present and on their position on the parent dibenzodioxins ring structure. It was not until 1957 when a German dermatologist, Dr. Karl Schultz from the University of Hamburg, detected a severe form of acne named *chloracne* on the skin of many workers of a nearby chemical plant.[90] Dr. Shultz and his colleges started further research on rabbits and determined later that TCDD caused liver damage and death in rabbits. At about the same time in the United States, the Food and Drug Administration was trying to find the cause of death of millions of chickens. The answer finally came in 1966, when it was determined that TCDD was one of the contaminants responsible for the outbreak of disease among the chickens.[91]

On July 10th, 1976 the town of Seveso in Italy was contaminated with TCDD due to an explosion at a chemical plant's reactor, spreading a cloud of vapor poisoning over 38,000 people.[92] The reactor toxic contents were blown up when a failure in the cooling system occurred, raising the temperature to an alarming heat that caused it to burst, releasing 250 grams of dioxin into the atmosphere. That cloud of vapor consisted primarily of 2,4,5-trichlorophenol but also contained the extremely toxic reaction by-product 2,3,7,8-tetrachlorodibenzeno-p-dioxin.[93] As a result, 3,000 rabbits and chicken died almost immediately after the accident and 183 people, mostly children, were diagnosed with chloracne. Another unfortunate episode of dioxin contamination happened near the city of Niagara Falls, New York, in a site known as Love Canal.

A chemical company by the name of Hooker Chemicals began dumping waste in the Canal around 1930. Among the chemical waste dumped into the canal was a heavy amount of chlorinated hydrocarbons and ash from their incinerators. Ac-

---

88 A. Hay, The Chemical Scythe: Lessons of 2,4,5-T and Dioxin 30-35 (University of Leeds, Department of Chemical Pathology, 1984).

89 C. Rappe, H.R. Buser & H.P. Bosshardt, *Dioxins, dibenzofurans and other Polyhalogenated aromatics: Production, use, formation and destruction*, 320 Annals of the New York Academy of Sciences 1-18 (1979).

90 M. Gough, Dioxin, Agent Orange 29-33 (Plenum Press 1978).

91 D. Firestone, *Etiology of Chick Edema Disease*, Environmental Health Perspective 58-66.

92 P. Mastoiacovo, et al., *Birth Defects in the Seveso Area after TCDD Contamination*, 159 J. of the Am. Med. Ass'n. 1668-72 (1988).

93 A. Hay, *Toxic Cloud over Seveso*, 262 Nature 636-38 (1976).

cording to Hooker's records, some 21,800 tons of waste were released into the Canal in ten years[94] of operation, among which EPA identified almost 200 different chemical substances. It is estimated that the amount of dioxin dumped into the Canal was 20 times higher than that following the accident in Seveso. Complaints by local residents arose in the spring of 1978 and signs of potential hazards began to appear, according to Dr. David Axelrod, Health Commissioner for the State of New York at the time.[95] On April 15th, 1978 the Department of Health of New York issued a public note suggesting that Love Canal residents were facing serious health hazards. This was confirmed a month later by a team of scientists from EPA, when they analyzed and found toxic vapor in the area. In August of the same year, 236 families were evacuated from the area near the Canal and an eighty-foot high fence was placed around the site. Later 710 other families were also relocated because the area was contaminated, posing a serious threat to their health.[96] Some of the problems faced by the inhabitants of Love Canal were an increase in spontaneous abortions among women, incidence of malformed children, and genetic aberrations such as breaks, marker chromosomes, and ring chromosomes.[97] It is proven that TCDD causes cancer to animals. The tumor is most likely to develop in the liver, lung and thyroid.[98]

Air currents easily transport dioxins and furans by particles to which they are bound. Particles of soil are rich in organic carbon, and dioxin and furans have great affinity for organic carbon. As a consequence, they bind easily to these particles. In this manner these particles can be lifted and carried away by wind or water to great distances. Dioxin and furans have both been found in fat tissue of Arctic seals, thousand of miles away from where they have originated.[99] High levels of dioxin and furans are proven to be responsible for birth defects, child growth retardation, reduced level of male reproductive hormones, diabetes, and cancer.[100] The International Agency for Research on Cancer has classified dioxins as a human carcinogen.

Information above mentioned excerpted from:

1) Dictionary of Environmentally Important Chemicals
D.C. Ayres and D.G. Hellier

---

[94] D. Axelrod, *Chlorinated Hydrocarbons (US Love Canal): Case Studies of Selected Area-wide Environmental Exposures*, Address Before The National Academy of Sciences Workshop on Plans for Clinical and Epidemiological Follow-up after Area-Wide Chemical Contamination (Mar. 17th, 1980).

[95] Id.

[96] M. Brown, Laying Waste (Pentheon Books 1980).

[97] Biogenics Corporation, Pilot cytogenetic study of the residents of Love Canal (May 14th, 1980).

[98] J.R. Allen, J.J. Lalich & J.P. Van Miller, *Increased incidence of neoplasma in rats exposed to low levels of 2,3,7,8-tetrachlorodibenzo-p-dioxin*, 9 Chemosphere 537-44 (1977).

[99] M. Oehme et al., *Presence of Polychlorinated Dibenzo-p-dioxins, Dibenzofurans, and Pesticides in Arctic Seal from Spitzbergen*, 17 Chemosphere 1291-1300 (1988).

[100] M. Fingerhut et al., *Cancer Mortality in Workers Exposed to 2,3,7,8-TCDD*, 324 New England J. of Med. 212 (1991).

Queen Mary and Westfield College, University of London, UK
Fitzroy Dearborn Publishers, London. 1998.

2) Dioxins and Furans: Questions and Answers.
Todd Paddock. Academy of Natural Sciences
Philadelphia, PA. 1989.

3) Assessment Report prepared for the IPCS on: DDT, Aldrin, Dieldrin, Endrin, Chlordane, Heptachlor, Hexachlorobenzene, Mirex, Toxaphene, Polychlorinated Biphenyls, Dioxins, and Furans.
Prepared by: L. Ritter, K.R. Solomon, J. Forget.
Canadian Network of Toxicology Centers—December 1995.

4) United Nations Environment Programme—UNEP Chemicals: Official Website: http://www.chem.unep.ch/pops/indxhtms/asses6.html.

D. Examples of the Effects of POPs on Health and the Environment

During the Second World War, the world experienced a shift away from the use of inorganic chemicals as pesticides.[101] The old practice of using simpler insecticides derived from naturally occurring minerals such as arsenic, copper, zinc, manganese, and lead gave way to the widespread use of carbon-based pesticides. These synthesized pesticides exhibited far greater potency than those derived from natural plant sources such as nicotine sulfate (extracted from tobacco leaves) or pyrethrum (extracted from chrysanthemum flowers).

Until 1650, the average life expectancy for humans was rarely more than 25 years.[102] This number has increased dramatically, and in some parts of the world, the average life expectancy has jumped to 80 years.[103] The development of penicillin initiated an era of wide-spectrum antibiotics, which became powerful agents for the treatment of a wide array of diseases.[104] This, in turn, enabled people to become more healthy and productive. Advances in science and the chemical industry have together contributed significantly to the development of successful treatment for many other types of diseases such as tuberculosis, rheumatic fever, syphilis, poliomyelitis, measles, rubella, influenza, and some forms of cancer, to name but a few.[105] However, these advances in human health have not been without cost. In the process of developing better medicines and pesticides, man has created environmental hazards in the form of air, water, and soil pollution,[106] which have had a deleterious effect upon human health. Many morphological abnormalities and certain cancers may be largely accounted for by exposure to cer-

---

101 JOHN WARGO, OUR CHILDEN'S TOXIC LEGACY: HOW SCIENCE AND LAW FAIL TO PROTECT US FROM PESTICIDES (2d ed. 1998).
102 K.D. FISHER ET AL., THE SCIENCE OF LIFE: CONTRIBUTIONS OF BIOLOGY TO HUMAN WELFARE (Plenum/Rosetta ed., 1972).
103 *See id.*
104 *See id.*
105 *Id.*
106 GLOBAL EFFECTS OF ENVIRONMENTAL POLLUTION (S.F. Singer ed., 1970).

Chapter 2

tain pollutants that we have released into the environment.[107] Our health has always had a close relationship with the environment. We are in a constant state of interaction with our surroundings, and our survival depends on the delicate balance of what we take from the environment and what we release into it. The ever-increasing spectrum of new chemicals created in laboratories reaches our bodies through the water we drink, the food we eat, and the air we breathe.

The term "pollutant" normally refers to synthetic chemical compounds, but there are four other important categories of chemical pollutants:[108]

a) Natural chemicals in excess, such as nitrates that are normally found in the diet; high consumption of nitrates can lead to blood disorder among children.

b) Toxic chemicals found in fungi, plants or crop products, such as aflatoxins and cycasins.

c) Organic and inorganic mixtures, which includes air and water pollutants.

d) Synthetic chemicals, such as pesticides, insecticides, herbicides, fungicides, food additives, fuel additives, household chemicals, industrial chemicals, and pharmaceuticals.

What interests us most is the fourth category, for synthetic chemicals possess the power to alter vital biological processes. This can occur when synthetic chemicals destroy the enzymes present in our bodies, or alter them in such a way as to prevent their normal functioning. Some chemical pollutants block the enzymes required for the oxidation process, thereby interfering with the way our bodies receive energy. Chemical alteration of other enzymes prevents the normal functioning of vital organs, and in some cases, the alterations are severe enough to cause tissues to become malignant.

Very little was known about the effects of POPs on human health and the environment until the Sixties. However, since then the scientific community has put together a wide array of evidence on POPs. This evidence has played a key role in the development of measures to prevent further damage by these toxic substances. These efforts have enabled us to better understand some of the morphological and behavioral abnormalities observed in fish, birds, and mammals worldwide. In many cases, the evidence points primarily to POPs present in the environment as the cause of these abnormalities.

1. Some Reported POPs Effects in the United States

The abundant use of anthropogenic chemical compounds has spread worldwide over the last fifty years. Over the years, effects of the use of these chemicals on the environment, human health and fauna has been shown. The production of

---

[107] H. E. Stockinger, *The Spectre of Today's Environmental Pollution*, 30 AMERICAN IND. HYG. ASSOC. J. 195 (1969).

[108] THE SCIENCE OF LIFE: CONTRIBUTIONS OF BIOLOGY TO HUMAN WELFARE 243-294 (K.D. Fisher, & A.U. Nixon eds., 1972).

synthetic pesticides in the United States alone went from 124,259,000 pounds in 1947 to 637,666,000 pounds in 1960, resulting in over a quarter of a billion dollars in revenue to the manufacturers.[109] In 1990, the U.S. manufactured and exported over 465 million pounds of different kinds of pesticides. From this production, 52 million pounds were classified as banned, restricted or prohibited for use in America by USDA or EPA, but yet they were still produced and exported to other parts of the world, according to U.S. Customs data.[110]

Many of the effects on human health and the environment were hard to detect, due to the lack of technology. However, in 1952, a method of chemical analysis called paper chromatography was introduced. This method was particularly useful in the detection of various chlorinated compounds. Ten years later gas chromatography brought a tremendous advancement in the detection of pesticides. This method could detect compounds in the range of 0.02 to 0.2 parts per million.[111]

As the years went by, other advances helped to improve our understanding of how pesticides move through the environment and affect our health. In 1970, refinements in mass spectrometry created a revolution in environmental chemistry. Mass spectrometry allowed more accurate detection of chemicals in the environment which, in turn, enabled policy makers to establish legislation which set absolute levels of tolerance for certain chemicals.

In a world of global trade it is almost impossible to avoid contamination due to the heavy use of pesticides by nations worldwide. Unfortunately, the knowledge of patterns of international pesticide exchange is quite limited, which creates a complex problem that is virtually impossible to monitor. The world's largest exporter of pesticides is Germany, followed by the United States and the United Kingdom.[112] In 1988 the United States imported 42,000 pounds of beef from Honduras, which tests showed to contain heavy levels of chlordane and heptachlor, both of which were prohibited for use and banned in the U.S. since 1983. Agricultural imports to the U.S. usually originate in developing countries. In 1979, nearly 7% of all agricultural imports to the United States came from Nicaragua, El Salvador, Honduras and Guatemala. These countries use banned pesticides extensively on the crops, which are exported to the U.S., including DDT, endrin, toxaphene, and dieldrin.

2. In the Arctic Region

Several groups of people in the Arctic are highly exposed to contaminants, in particular to POPs, which are carried by air currents or water and accumulate in

---

[109] Report, The United States Department of Agriculture, Commodity Stabilization Service, *The Pesticide Situation for 1960-61* 1-23 (July 1961).

[110] C. Smith, & S.L. Beckman, Export of Pesticides from the U.S. Ports in 1990: Focus on Restricted Pesticide Export. Senate Committee on Agriculture, Nutrition and Forestry, Hearing on the circle of poison, 2.

[111] L.C. Mitchel, *A New Indicator for the Detection of Chlorinated Pesticides on the Paper Chromatograph*, 35 Assoc. Off. Agric. Chemistry J. 928 (1952).

[112] United Nations Food and Agriculture Organization, Trade Yearbook 311-13 (1990).

Chapter 2

animals that are used as traditional foods. The Inuit, traditional inhabitants of the Arctic, are indigenous people who are closely linked to local resources, and who also depend heavily on wildlife and fish as sources of protein in their diet. Due to its location, severe weather conditions, and seasonal variations in light, the Arctic is a potential storage sink for contaminants. Strong south-to-north airflows originating in Europe bring POPs and other contaminants to the Arctic, especially in the winter. Another major way contaminants reach the Arctic is through ocean water due to the movement of marine currents. Most industrial pollutants released into the atmosphere on the Kola Peninsula (in the northwestern Russia), eastern Finland, and Norilsk, for example, will find their final destination in the Arctic. Dioxins and furans from smelters in Norway have also been identified within the Arctic.

The Island of Broughton, in Canada, hosts an Inuit population of 450 people, who live some 70 miles north of the Arctic Circle. They have all been contaminated with extremely high levels of PCBs. The Inuit women who live in northern Quebec have also shown higher levels of DDE and PCBs than Canadian women who live in the southern part of the country. This occurs mainly because the Inuit people use animals high on the food web as a daily part of their diet. This contamination is passed onto the developing fetus and to breastfeeding children, causing serious threats to Inuit children, including neurological disorders in babies and cancer. It has been shown that the concentration of contaminants in Inuit women's breast milk is five to ten times higher than in other women in the rest of Canada. These women are more vulnerable to reproductive effects and increased risk of breast cancer. The saddest part of the story is that none of the POPs are manufactured nor used by the Inuit people, and yet they are suffering the highest risk, along with other individuals who live in the Arctic region.

In 1991, ministers from the eight Arctic countries[113] got together to examine the problem of anthropogenic pollution in the region and its effects on human health and the environment. As a result of the meeting they established the Arctic Monitoring and Assessment Program ("AMAP"), with the objective to collect relevant data and scientific literature concerning the problems affecting their population and the environment. Under the Arctic Environmental Protection Strategy ("AEPS"), programs were established for monitoring the movements and effects of pollution. The programs involved the efforts of some 400 scientists, and resulted in an extensive assessment report, called "The AMAP Assessment Report: Arctic Pollution Issues."[114] The report concluded that the level of contamination in people who use the Arctic food webs is 10 to 20 times higher than in most regions of the world. This is of great concern because indigenous peoples of the Arctic are the ones who rely mostly on traditional diets and, therefore, are more exposed to high levels of contamination. This is due to the fact that POPs bioaccumulate in the fatty tissues. Most marine mammals and fish

---

113 Arctic Countries are: Canada, Denmark, Greenland, Finland, Iceland, Norway, Sweden, Russia and the United States.
114 AMAP, *at* http://www.amap.no/assess/soaer0.htm (last visited Apr. 5, 2002).

in the Arctic, for example, are rich in polyunsaturated fat, which is used as a traditional food. The majority of the people who were tested showed high levels of almost all of the twelve persistent organic pollutants herein detailed.[115] Reindeer and caribou are typical herbivorous mammals of the Arctic that eat ground vegetation. Due to atmospheric deposition of chemicals on their grazing areas, many animals have shown high levels of hexachlorobenzene and hexachlorocyclohexane in their liver. High levels in the concentration of PCBs, hexachlorobenzene and hexachlorocyclohexane are found along the shore on the Norwegian coast, river mouths along the Russian coast, and around Svalbard. The cause of this concentration of contaminants is probably industrial and municipal effluents and runoffs.

---

115 Information obtained from the AMAP's web site: AMAP, *at* http://www.amap.no/assess/soaer12.htm (last visited Apr. 5, 2002).

CHAPTER 3

BACKGROUND ON GLOBAL ACTION ON HAZARDOUS CHEMICAL SUBSTANCES PRIOR TO THE UNITED NATIONS STOCKHOLM CONVENTION ON PERSISTENT ORGANIC POLLUTANTS

A Political Setting: The Context and National Legal Measures

1. The Nature of the Problem: Growing Awareness

In the 1960s the developing countries experienced a burst of growth in agricultural practices called the "green revolution," by an increase in food grain production per acre due to the use of high-yielding varieties, pesticides, and improved management practices.[1] This was in response to the growing demand of food due to rapid population growth in developing countries.[2] During this "green revolution" phase from 1970 to 1994, food grain production increased by approximately 780 million tons, or 79 per cent.[3] The only problem is that these seed varieties were less pest resistant and required more chemical use to control a pest outbreak, resulted in increased pesticide use.[4] The most conservative statistics show that global pesticide use has doubled in every decade between 1945 and 1985, in particular by developing countries.[5] It is estimated that each year four billion pounds of pesticides are produced in the world by developed countries, of which 75% of this production is for agricultural purposes in developing nations.[6] These chemicals are shipped all over the world. Since the variety of languages and cultures in developing countries makes communication in a common language difficult labeling and information about the dangers of that chemical substance, and handling instructions, for example, are nonexistent.[7] With the increase of world trade in chemicals between the sixties and seventies,[8] people in many countries were alerted by scientific reports being published about the risks of using hazardous pesticides. Popular awareness of this danger grew virtually overnight with the publication of Rachel Carson's *Silent Spring* in 1962, which helped to spark the postwar environmental movement.[9]

---

1   J.H. WEAVER, ET AL., ACHIEVING BROAD-BASED SUSTAINABLE DEVELOPMENT: GOVERNANCE, ENVIRONMENT, AND GROWTH WITH EQUITY 140–149 (1997).
2   *See id.*, at 146.
3   *See id.*, at 146; *See also* Report, The World Bank, *World Development Report* 168 (1994).
4   Faith Halter, *Regulating Information Exchange and International Trade in Pesticide and Other Toxic Substances to Meet the Needs of Developing Countries*, 12 COLUMBIA J. ENVTL L. 1, 3-4 (1987).
5   G. SPETH, *Environmental Pollution* Earth '88: Changing Geographic Perspectives, National Geographic Society. Washington, DC.
6   58 Fed. Reg. 9062-63 (1993).
7   William C. Kalmabach, III, Note, *International Labeling Requirements for the Export of Hazardous Chemicals: A Developing Nations Perspective*, 19 L. & POLICY INT'L BUSINESS 811-17 (1987).
8   OECD, ECONOMIC ASPECTS OF INTERNATIONAL CHEMICAL CONTROL (1983).
9   RACHEL CARSON, SILENT SPRING, (1962).

The significant increase of pesticide poisoning in the developing nations, while affecting millions, caused some 220,000 documented deaths in the past twenty years, according to one study.[10] The use of chemical substances for agricultural pest control purposes in developing countries is also a major cause of environmental damage, such as water contamination, resulting in fish, wildlife, and human health poisoning and death.[11]

Despite these health concerns, the movement of pesticides around the world generates billion of dollars every year. Most of the pesticides manufactured and exported from developed countries have been banned in the industrialized world. For example, DDT has been banned in the United States since 1970, but U.S. manufacturers continue to export DDT to Nicaragua and Guatemala.[12] Aldrin, dieldrin and chlordane have also been banned in the United States; however, some American chemical companies still manufacture and export large amounts to the developing world.[13]

The drafting of Agenda 21 and its adoption in Rio in 1992 was the first multilateral document to acknowledge the full breadth of the problem of toxic substances commonly used throughout the world in agriculture and other daily activities. What led up to the inclusion of Articles 17 and 19 in Agenda 21 had begun decades before, especially with the groundbreaking publication of *Silent Spring*, which was widely read beyond the scientific community.

2. National Regulation of Hazardous Chemicals

Most developed states moved to ban POPs domestically during the 1970's through national legislation or regulation, other states, primarily developing countries, are just beginning this same process. The following is a brief selection of notable efforts under selected national legal systems to reduce or eliminate POPs domestically. While most developed states moved to ban POPs domestically during the 1970s–1990s through national legislation or regulation, unfortunately, many developing nations still do not have the expertise or infrastructure to ensure the safe use, transportation, and disposal of pesticides and other hazardous substances.[14]

Within individual countries, people began to notice cases of pesticide poisoning, beginning with the publication of *Silent Spring*. Domestic legislation and regulations were adopted in some states long before any international movement began to address these problems. Common problems in developed as well

---

10 M.B. Baender, *Pesticides and Precaution: The Bamako Convention as a Model for an International Convention on Pesticides Regulation*, 24 N.Y.U. J. INT'L L. & POL 557– 60 (1991).

11 *See id.*, at 562.

12 D. WEIR, & M. SHAPIRO, CIRCLE OF POISON: PESTICIDES AND PEOPLE IN A HUNGRY WORLD 86-87 (Institute for Food and Development Policy ed., 1981).

13 *See* James Colopy, *Poisoning the Developing World: The Exportation of Unregistered and Severely Restricted Pesticides from the United States*, 13 UCLA J. ENVTL. L. & POL 167, 173 (1994-95).

14 *See id.*

as in developing countries at that time resulted from poor management of chemicals; lack of public awareness; primitive work conditions precluding the safe use of pesticides; lack of available protective clothing; poor storage facilities; the sheer quantity of pesticides in use in agricultural areas; and the danger of agricultural pesticides leaching into the groundwater. In addition, workers today in developing countries often live near pesticide storage places, which further exposes them and their families to the effects of the hazardous chemicals. Tropical weather increases the likelihood that workers will not wear protective clothing or gear, even if it is provided, due to discomfort and extreme heat.

In response, countries, such as Sweden, began to regulate the use and production of hazardous chemicals within their own countries. These developments, described in summary below, may be compared with similar efforts in key developing countries, such as Brazil.

2.1 Sweden

Sweden has a long tradition of an active and sophisticated chemicals policy. In 1985 the Chemicals Commission proposed a draft to the Parliament, which was approved as the "Chemical Products Act." At the same time, the Chemicals Commission also proposed the creation of a new agency to take care exclusively of chemical related issues, called the Chemicals Inspectorate, which is a supervisory authority under the Swedish Minister of the Environment.

Sweden has a very strong regulatory scheme on persistent organic pollutants. Aldrin and dieldrin were both banned in Sweden in 1969, hexachlorobenzene was voluntarily withdrawn from the market in the early seventies, and PCBs were regulated in the same time. DDT was banned in 1975, and some of the other persistent organic pollutants were never used in Sweden. Since the early eighties, the Swedish government has rigorously regulated all emissions of furans by requiring special filters on all smelters. Also air quality monitoring is performed by independent bodies and by governmental agencies, and the results must match. The operating permit is withdrawn if the industry does not comply with the standards set by the Chemical Inspectorate.

Later, on May 9th, 1996, the Swedish government undertook a restructuring of the Chemicals Commission. The Chemicals Policy Committee was established to review the old chemical policy and propose changes and more up-to-date chemical policy for the future. This was prompted by the Esbjerg Declaration, adopted at the 4th International Conference on the Protection of the North Sea where it was agreed by Nordic countries that all discharges of hazardous substances in the North Sea would be phased out by 2020.

Between 1997 and 1998, the Swedish government, who desired Sweden to be a driving force and a model of ecologically sound management of chemicals to the world, introduced major changes. In proposing these changes fifteen environmental quality goals were adopted by the parliament that had to be achieved in one generation. One of the goals of the newly established ecocycle[15] strategy is a "non-toxic environment," meaning the desire to have an environment free from man-made chemical substances. The Chemicals Policy Committee proposed that by the year 2007 all products, including chemicals, used in Sweden should be free from any substances that are persistent, bioaccumulate, or pose serious or irreversible adverse effects on human health or the environment.[16] It also established that by 2012 all production process of these substances should be phased out.

In 1997, the Swedish government presented to the parliament the Statement of Government Policy, stating a model for ecological sustainability, meaning "united all measures to meet the needs of the present without compromising the ability of future generations to meet their own needs." In the same year parliament approved the Swedish Economic Bill and in 1998 the Budget Bill in which it was established three main objectives to be sought in order to achieve this so-called "ecologically sustainable development":

a) protection of the environment;

b) sustainable supplies; and

c) efficient use of energy and other national resources.

Under the first objective of "protection of the environment," the government wishes to control emissions of pollutants in such a way that it will not damage humans or the environment, and it will not exceed nature's capacity for absorbing or breaking them down. These targets are to be met by a) cooperation between government, industry and consumers, b) new directions for the development of new products, c) widespread of information to create public awareness and educate the consumer, d) international cooperation, and e) stronger national laws.

The Swedish environmental legislation was reformed. A new Environmental Code[17] came into force on January 1st, 1999, aiming to promote sustainable development that ensures a healthy environment for both current and future generations. The scope of the new legislation is to protect human health and the environment against damage and nuisance, preserve biological diversity, promote reuse and recycling practices, and safe-guard long-term management of resources from an ecological, social, cultural, and socio-economic point of view.[18] It is interesting to note that before the adoption of the Stockholm Convention

---

15   **Ecocycle** strategy is a government program established by the Commission on Ecologically Sustainable Development in Sweden, in which natural cycles must be protected, and all man-made substances should not occur in nature. http://www.regeringen.se.

16   *Interim targets for a Non-Toxic Environment decided by the Swedish Parliament*, at http://www.internat.environ.se/documents/objetctiv/objdoc/obj12.html (last visited on Feb. 16, 2002).

17   Swedish Environmental Code came into force on January 1st, 1999. See http://www.internat.environ.se/documents/legal/code/codedoc/code.htm (last visited on Feb. 16, 2002).

18   *See id.*, at Ch. 1 (Objectives).

on POPs, Sweden selected for their national Environmental Code the same goal, namely to protect human health and the environment. All fifteen environmental laws that regulated various fields of the environment were combined in the new Environmental Code. These are: the Natural Resources Act, the Environmental Protection Act, the Dumping Waste in Water Act, the Fuels Act, the Agricultural Land Management Act, the Public Cleansing Act, the Health Protection Act, the Water Act, the Pesticides Act, the Chemical Products Act, the Environmental Damage Act, the Nature Conservancy Act, the Biological Pesticides Act, the Genetically Modified Organisms Act, and the Flora and Fauna Act. The Swedish Environmental Code in Chapter 14 has a section on "Chemical products and biotechnical organisms" which sets the rules for approving imports of chemical substances. Any pesticides, whether chemical or biological, may be imported only from European Union states, and even then subject to approval of the Chemical Inspectorate at five-year renewable intervals. The criterion for this approval is safety for health and the environment. Chemical products and biotechnical organisms that are not subject to permit or approval requirements may also have to submit information to the Chemical Inspectorate if new information becomes available suggesting the substance may have carcinogenic, mutagenic or reproduction-toxic properties that affect its labeling or classification.[19]

Domestically, Sweden has its own products register in Chapter 14, Section 10 of the Environmental Code. Professional manufacture or import into Sweden of chemical products in quantities over 100 kilos (200 pounds) must be reported to the Product Register, similar to the prior informed consent procedure internally within one country. "Manufacturer" for this purpose includes anyone who repackages, packages, or changes the name of a chemical product for further transfer of the product. The Chemical Inspectorate maintains the Register.[20]

Sweden has been a world leader in the field of POPs in the 1990s, both in its domestic legislation and in its role in the negotiation of the Stockholm Convention. Nothing pertaining to POPs was on the agenda of the 1972 Stockholm Conference on the Human Environment. Since that time, however, as awareness of the risks of POPs grew, Sweden has taken direct action and established laws, regulations, and enforcement mechanisms to regulate the production and use of POPs and limit imports of such substances. Classification and labeling of chemical products to indicate risks are based on the product's properties and internationally adopted criteria. Sweden has developed its own system to inform the user of the risks and the precautions that should be taken, which permits consumers to understand through clear language and choose their purchases accordingly. Sanctions for those who violate the Environmental Code are monetary fines levied directly by the government when a violation has occurred.

---

19 *See id., supra* note 202, at § 16 (Environmental Code of Sweden).
20 Kemikalieinspektionens föreskrifter—KIFS (Swedish Chemical Inspectorate) Ch. 7 (1998). http://www.kemi.se/default_eng.htm. Site visited on April 8, 2002.

2.2 Brazil

CONAMA, the National Council on the Environment, is the entity charged with establishing patterns of environmental quality, regulation of environmental zoning, and the monitoring of all activities that potentially may be polluting.[21] The Council votes on rules and standards. Members of the Council are nine government officials appointed by the President of the Republic. It establishes the norms and criteria for environmental permitting. These are administrative rules and procedures. The majority of rules pertaining to POPs in Brazil come from CONAMA rather than from the Ministry for the Environment or Federal laws from the Brazilian Congress. Laws enacted by the Congress, of course, would be stronger than administrative regulations, but in the case of POPs, CONAMA is the major source rather than the legislature. In the future when Brazil ratifies the POPs Convention, then Federal laws will be enacted to implement the treaty.

The following lists those laws, which are relevant to POPs in Brazil:

a) Federal Law 19/1981—Prohibits any procedure that will produce PCBs. Prohibits PCB use and trade. Establishes that existing electrical transformers can only be replaced by others that do not use PCB.

b) Ordinance no 001/10, June 1983. Set rules for the management, storage, and transport of PCBs and residues contaminated with PCBs.

c) Law 329/85 from the Ministry of Agriculture: Prohibits use, commerce, and distribution of organochlorinated pesticides (aldrin, DDT, mirex, heptchlor, endrin, toxaphene, HCB) for farming and cattle-raising purposes.

d) Resolution 20/1986 from the National Environmental Council (CONAMA). Establishes percentage doses of potential hazardous substances in water discharge of organochlorinated compounds (DDT, mirex, heptachlor, dieldrin, endrin, chlordane, aldrin, toxaphene, and PCBs).

e) Resolution 6/1988 from the National Environmental Council (CONAMA). Establishes a control mechanism for industrial discharges of organochlorinated compounds (DDT, HCB, and PCBs).

f) Resolution no. 36, January 19th, 1992 from the Ministry of Health. Establishes norms and standard patterns of water potability for human consumption.

g) Resolution no. 91, November 13th, 1992 from the Ministry of Agriculture. Prohibits any registration, production, import, export, commerce, and use of derivates of hexachloropentadiene.

h) Resolution 19/1994 from the National Environmental Council (CONAMA). Authorizes the export of hazardous residues contaminated with PCBs.

---

21 Paulo Affonso Leme Machado, Direito Ambiental Brasileiro 127-129 (8ª Edição, 2000).

i) Resolution 23, December 12th, 1996 from the National Environmental Council (CONAMA). Regulates import of organochlorinated residues in general in accordance with the Basel Convention requirements.

j) Law 204/1997 from the Ministry of Transportation. Sets rules for the transport by road or train of hazardous products such as DDT, HCB, mirex, heptachlor, endrin, chlordane, aldrin, and PCBs.

k) Ordinance no. 11, January 8th, 1998 from the Ministry of Health. Bans organochlorinated products from the list of Hazardous Substance that can be authorized for use in farming and for sanitary purposes.

l) Ordinances 8, 9, and 10, May 18th, 1999 from the Ministry of Agriculture. Establishes monitoring program for the production and export of citric pulp left from extraction of juice, used for animal feed.

2.3 Bilateral (U.S.-Canada)

One binational effort to find a solution for POPs is the United States and Canada's program for a "Binational Strategy for the Virtual Elimination of Persistent Toxic Substances in the Great Lakes" in April 7th, 1997.[22] This agreement is designed around an appendix, which is a list of "Persistent Toxic Substances" in two levels, both of which contain POPs chemicals. Level I substances are those previously named in lists relevant to pollution of the Great Lakes for their bioaccumulation. Level II substances are substances for which one country or the other has grounds to indicate its persistence in the environment, potential for bioaccumulation and toxicity. In Level I, aldrin, dieldrin, chlordane, DDT, hexachlorobenzene, mirex, PCBs, dioxins and furans are also POPs. Level II, endrin, heptachlor are also POPs, so they are found in the Stockholm Convention as well as in the Great Lakes Binational Toxics Strategy. The Great Lakes hold 18% of the world's supply of fresh water, and 47% of the 33 million people living around the lakes draw their drinking water from the lakes. The lakes are vital to many fish and wildlife species. Pollution from POPs endangering fish, wildlife and humans was first noted in the 1970s. Under the Great Lakes agreement, all discharge of any and all POPs is to be eliminated immediately, beginning with a plan to accomplish this objective on both sides of the border between Canada and the U.S.

B. "Soft Law" Measures to Regulate Hazardous Chemical Substances on the International Level

Tentative efforts had begun in the United Nations system to create the basis for a system of control or regulation, without noticeable success. One such early attempt was the listing of suspected or known hazardous chemical substances in the International Register of Potentially Toxic Chemicals, which merely made

---

22 Stated at the Washington Declaration on Protection of the Marine Environment from Land-Based Activitieshttp://irptc.unep.ch/pops (1995).

data available, but did not require any action by any member state of the UN. This inventory was an outgrowth of the first UN environmental conference in Stockholm in 1972.[23]

Other attempts were made to establish soft law documents in the two decades leading up to the Rio conference on environment and development, including two attempts to regulate trade: the 1985 International Code of Conduct on the Distribution and Use of Pesticides by FAO and the 1987 London Guidelines for the Exchange of Information on Chemicals in International Trade by UNEP. Until Agenda 21, however, these documents remained less than universal statements, since they were made by specialized agencies of the United Nations and not binding on all member states of the UN.

These FAO and UNEP documents were merely voluntary and had little or nothing to do with the developing world, which had no means of implementing the costly suggestions contained in these codes. In addition, developing countries heavily dependent on agriculture and chemical pest control had no affordable alternatives, even those that might be available to industrialized countries with research capabilities and corporate agribusiness interests.

Only when the developed world had taken steps to eliminate many of the most hazardous chemicals in agriculture and in public health campaigns for insect control, for example, did their attention turn to the problem of agricultural products being imported from developing countries that contained the residue of the same hazardous chemicals no longer in use in the importing states. Following UNCED, three other developments occurred which lead to a new international law on POPs: the creation of the Intergovernmental Forum on Chemical Safety (IFCS); UNEP's Code of Ethics; and creation of the Inter-Organization Programme on the Sound Management of Chemicals (IOMC).

An analysis of each of these voluntary codes and entities is useful, however, to understand the growth of scientific understanding of the threat posed to human health by the use of "potentially toxic chemicals" and the early stages in translating this scientific evidence into policy and law.

After 1992, these early soft law documents, including: Agenda 21; voluntary guidelines; and the scientific work of the organizational entities created in the international community, such as the IRPTC, IFCS, and IOMC, formed the basis for drafting treaties which become "hard law" with binding legal obligations on states.

1. The International Register of Potentially Toxic Chemicals—IRPTC (1976)

The United Nations Environmental Program established IRPTC as a databank of potentially toxic chemicals in 1976, following a recommendation of the 1972

---

23   Stockholm Conference on the Human Environment (1972).

United Nations Conference on the Human Environment,[24] held in Stockholm. Any state or NGO could consult the IRPTC to see what substances were listed as "potentially toxic." For the first time, industrialized and developing nations met to delineate the role of human beings and their activities in order to preserve a healthy and productive environment.[25] A centralized international body created an institution to gather information on hazardous chemicals and make it available to every nation. IRPTC is a comprehensive register of which chemicals are being produced, where, by which company, in what quantities, for what purpose, and their chemical properties, toxicology and effects when released into the environment.[26] This international register is a valuable tool to help nations around the world to make better use of existing global resources and share information regarding the use, effects, potential hazards, biochemical properties and handling of toxic substances.[27] It is also intended to help policy makers when setting goals on public safety and environmental health policy at local, regional and international levels. Since 1976, IRPTC has been used widely by all states; perhaps today it is of greatest use to developing countries that would otherwise be unable to collect such information.

In 1989 UNEP's Governing Council revised the goals of IRPTC into five categories:[28]

a) To make it easier to obtain the existing information on production, distribution, release, disposal and adverse effects of chemicals;

b) To identify the important gaps in our knowledge of the effects of chemicals and call attention to the need for research to fill those gaps;

c) To help identify potential hazards from chemicals and wastes and to improve awareness of dangers;

d) To provide information about national, regional and global policies, controls and recommendations on potentially toxic chemicals, and

e) To help implement policies for the exchange of information on chemicals in international trade.

In 1981 the Governing Council of UNEP broadened the mandate of the group working on the IRPTC to address the adverse effects of toxic chemical substances on humans and on the environment and called the office "UNEP Chemi-

---

24 *See* THE WORLD COMMISSION ON ENVIRONMENT AND DEVELOPMENT, OUR COMMON FUTURE (THE BRUNDTLAND REPORT) 177 (Oxford University Press 1987). "A lot of policy development work is needed. Nations urgently need to formulate and agree upon management policies for all environmentally reactive chemicals released into the atmosphere by human activities, particularly those that can influence the radiation balance on earth. Governments should initiate discussions leading to a convention on this matter. In if a convention on chemical containment policies cannot be implemented rapidly, governments should develop contingency strategies. . . ." *Id.*

25 United Nations System-Wide Earthwatch, *UNEP IRPTC (Chemicals) guidelines, at* http://www.unep.ch/earthw/Pdepche.htm (last visited Feb. 10, 2002).

26 UNEP– IRPTC, *at* http://www.chem.unep.ch/irptc/.

27 *See* Mr. James Willis, Director of the UNEP Chemicals, declaration, *at* http://www.unep.org/unep/program/hhwb/chemical/irptc/home.htm (last visited Apr. 5, 2002).

28 *See id. See also* UNEP IRPTC, *at* http://www.chem.unep.ch/irptc/ (last visited Apr. 5, 2002).

cals." This office, based in Geneva, continues to administer the IRPTC as a broad registry of chemicals and their toxicity, effects, and chemistry. Increased global demands from NGOs and governments for concrete action on chemical hazards led to requests for assistance to governments in building "chemical control capacity," meaning management of chemicals, their handling, disposal, and use. This function is provided by UNEP Chemicals. In addition, UNEP Chemicals led the work of promoting the Stockholm Convention, as explained below in Chapter 4.

IRPTC functions through a network of national and international organizations and industries, and has developed an extensive databank on chemical substances. This databank provides information on the distribution, release, disposal, and adverse effects of various substances on humans and the environment. It also provides a central library resource or legal backup center on existing national regulations for controlling toxic substances submitted by various states. Some of the chemicals listed in the IRPTC are also chemicals contained in the initial listing of the Stockholm Convention ("the dirty dozen").

2. The United Nations Food and Agriculture Organization International Code of Conduct on the Distribution and Use of Pesticides (1985)[29]

To face the problem of hazardous pesticide trade, marketing, and use in developing countries, the United Nations Food and Agriculture Organization (FAO) suggested the establishment of a set of rules that would serve as a guideline, on a voluntary basis, for governments worldwide.[30] This international response to the trade in dangerous chemical substances sent to the developing world was named the "International Code of Conduct on the Distribution and Use of Pesticides," adopted at the FAO Conference, 23rd Session, in November 28, 1985, through Resolution 10/85.[31]

The Code invites voluntary adherence to a set of rules with the objective of serving as a point of reference on the use, transport, and disposal for all nations, especially those that lack appropriate legal safeguards.[32] Many developing countries lack the infrastructure to register pesticides, to ensure their safe and effective use, to assess residues accurately, to put into practice labeling mechanisms, to safely distribute pesticides within the country, to promote risk evaluation, or to establish disposal methods in a sound and safe manner.[33] The equipment to assess chemical residues in agricultural products, for example, is prohibitively

---

29 The full text of the International Code of Conduct on the Distribution and Use of Pesticides can be found *at*: http://www.cepis.ops-oms.org/muwww/fulltext/toxicolo/codigo/codigo.html.
30 *See* Food and Agriculture Organization, International Code of Conduct for the Distribution and Use of Pesticides, Nov 25, 1985, U.N. Doc. M/R8130/E/5.86/1/5000 (1986).
31 FAO Conference Resolution 10/85, *at* http://www.foa.org/ag/agp/agpp/pesticid/Code/Annex.htm (last visited on Jan. 24, 2002).
32 *See id.* at art. 1(1.1) (Objectives of the Code of Conduct).
33 K.A. Goldberg, *Efforts to Prevent Misuse of Pesticides Exported to Developing Countries: Progressing Beyond Regulation and Notification*, 12 Ecology L. Q. 1025, 1030 (1985).

expensive and requires highly trained technical personnel in short supply in many parts of the world. In Africa, for instance, of all the countries that are members of the Southern African Development Community, only Tanzania, Mozambique and Zimbabwe have a legal framework to regulate pesticides.[34] The Code serves as a reference for any developing nation to freely consult until they have established their own regulatory framework for pesticide control.[35] The Code thus provides both capacity building and standardization by supplying a consistent, coherent body of data reflecting the best available science collected from around the world.

The Code was amended in 1989 to include Prior Informed Consent (PIC) mechanism.[36] PIC is defined as: "the principle that international shipment of a pesticide that is banned or severely restricted in order to protect human health or the environment should not proceed without agreement, where such agreement exists, or contrary to the decision of the designated national authority in the participating importing country."[37] This mechanism as part of the FAO Code of Conduct is distinct, of course, from the subsequent PIC Convention (see subsection D (5) below). The amended Code created a voluntary process for controlling exports of pesticides to the developing world. The amendment brought rules for exporting countries to benefit importing countries from information exchange and export notification for pesticides that were banned or severely restricted in the country of export.[38] In practice, this happens when a given country decides to ban or severely restrict the use or handling of a pesticide. The country must contact the FAO, who in turn will notify all other countries of that action through the International Register of Potentially Toxic Chemicals.[39] This enables participating nations to "assess the risks associated with the pesticides, and to make timely and informed decisions as to the importation and use of pesticides concerned, after taking into account local, public health, economic, environmental and administrative conditions."[40] Therefore, all banned or severely restricted pesticides in member countries become subject to the PIC procedure. Further, "no pesticide in these categories should be exported to an importing country participating in the PIC procedure contrary to that country's decision made in accordance with the FAO operational procedures for PIC."[41] In the event of a country's refusal of the import of a pesticide under the PIC proce-

---

34 See Report, USAID-Bureau for Africa, Environmental and Natural Resources Policy and Training Project & Winrock International Environmental Alliance, *Pesticides and the Agricultural Industry in Sub-Saharan Africa* iv (July 1994) (Executive Summary).
35 See FAO Code of Conduct, *supra* note 214, at art. 1.
36 FAO Resolution 6/89—Twenty-fifth Session of the FAO Conference (Rome, Nov. 11-29,) U.N. Doc. C/89/Rep.120 (1989).
37 See Code of Conduct, *supra* note 214, at art. 2 (Definitions).
38 See Code of Conduct, *supra* note 214, at art. 9.
39 See Code of Conduct, *supra* note 214, at art. 9.1 (Information exchange and prior informed consent).
40 See Code of Conduct, *supra* note 214, at art. 9.2.
41 See Code of Conduct, *supra* note 214, at art. 9.7.

dure, that country must halt any domestic production of that same pesticide. This approach was taken by FAO to avoid an importing country from using the PIC as a trade barrier in order to assist that country's domestic pesticide industry.[42]

As a mechanism of enforcement, the Code tries to promote "collaborative action" by all participating states,[43] suggesting that countries report to the FAO on their methods of compliance and progress.[44] The Code operates on a voluntary basis by participating countries. Consequently, it does not contain any mechanisms to enforce implementation, but rather relies upon collaborative efforts by all participating partners.

3. UNEP's London Guidelines for the Exchange of Information on Chemicals in International Trade (1987)

The Governing Council of the United Nations Environmental Program adopted the London Guidelines for the Exchange of Information on Chemicals in International Trade on June 17th, 1987.[45] The Guidelines were amended it in November 1989[46] to introduce further measures regarding information exchange on pesticides conforming to the PIC procedures noted above. At the June conference in 1987, the United Nations Environment Programme Governing Council recommended that UNEP launch a system for exchange of information between exporting and importing countries to supplement the London Guidelines. This system was referred to as the "prior informed consent procedure." The first draft revisions were completed in February 1989 and approved by the Governing Council in May 1989.[47] This procedure was subsequently followed by FAO in amending the Code of Conduct on the Distribution and Use of Pesticides in 1989, and became the embryo of the later Rotterdam Convention in 1998 on Prior Informed Consent.

Similar to the FAO Code of Conduct, UNEP's London Guidelines focused on the promotion of information exchange for the protection of human health and the environment.[48] The London Guidelines provide that the PIC procedure:

> means the procedure for formally obtaining and disseminating the decisions of importing countries as to whether they wish to receive future shipments of chemicals which have been banned or severely restricted. A specific procedure was established for selecting chemicals for initial implementation of the PIC procedures.

---

42 See Code of Conduct, *supra* note 214, at art. 9.8.2.
43 See Code of Conduct, *supra* note 214, at art. 12.1.
44 See Code of Conduct, *supra* note 214, at art. 12.6.
45 UNEP GOVERNING COUNCIL, 14th Sess., Agenda item 14/27, at 79, UN Doc. A/42/25 (1987).
46 Governing Council Decision 15/30, UN Doc. UNEP/GC.17/12, Annex II, at 17 (1989).
47 UNEP GOVERNING COUNCIL,. UN Doc. UNEP/PIC WG.2/L1/Rev.1 (May 25, 1989).
48 See *London Guidelines*, Introduction # 2, *at* http://irptc.unep.ch/ethics/english/longuien.htm (last visited Apr. 5, 2002).

These include chemicals which have been previously banned or severely restricted as well as certain pesticide formulations which are acutely toxic.[49]

The London Guidelines encourage exporting countries to use classification, labeling, and packaging requirements that are controlled by their own domestic market.[50] A significant innovation was introduced to insure that all instructions and warnings regarding pesticides have to appear in the language of the importing country.[51]

UNEP's London Guidelines were established parallel to the existing FAO Code of Conduct. Both provide guidance for the management of different classes of chemicals used in pesticides, and industry. The London Guidelines stress exchange of information for users of pesticides,[52] while the Code of Conduct affects manufacturers putting pesticides into the stream of commerce internationally.[53] Ultimately, FAO and UNEP joined together in a program to refine even further their common procedures for notification in the FAO/UNEP Joint Program. FAO is the lead agency for pesticides and UNEP Chemicals is the lead agency for other chemicals. The FAO/UNEP Secretariat established a list of chemicals subject to the Prior Informed Consent procedure, which includes twenty-seven chemicals, among which 7 are listed in the Stockholm Convention as POPs (aldrin, chlordane, dieldrin, DDT, heptachlor, hexachlorobenzene, and PCB).[54]

4. The United Nations Conference on the Environment and Development (UNCED) and Agenda 21 (1992)

Despite the creation of several international bodies to study and provide scientific data on POPs, it was not until 1992 that the absence of clear rules establishing acceptable global and regional standards for the handling of such toxic substances was first addressed. The delegates to the United Nations Conference on the Environment and Development ("UNCED" or the Earth Summit) recognized that chemical use is necessary to meet social and economic needs of the developing nations, but also poses a great danger to human health and the environment and that all needed to be done to ensure the sound management of chemicals worldwide.[55] Two documents emerged from the Earth Summit: the Rio Declaration and Agenda 21, the action plan for implementation of

---

49 *London Guidelines for the Exchange of Information on Chemicals in International Trade–Amended 1989*, Part I(1)(h), *at*: http://irptc.unep.ch/ethics/english/longuien.htm (last visited Apr. 5, 2002) (Definitions).
50 *See id.*, at art. 14.
51 *See id.*, at art. 13(d).
52 *See id.*, at no. 7 (Introduction to the Guidelines).
53 *See id*.
54 U.S. Environmental Protection Agency, Office of Pesticide Programs, *The Prior Informed Consent (PIC) Procedure: International "Right-to-Know," at* http://www.epa.gov/oppfead1/international/pic.htm (last visited on Feb.19, 2002).
55 *World Bank Report on the PIC Convention, at* http://www4.worldbank.org/legal/legen_int/legen_PIC.html (last visited on Jan. 23, 2002).

the declaration. Principle 14 of the Rio Declaration on Environment and Development says:

> States should effectively cooperate to discourage or prevent relocation and transfer to other States of any activities and substances that cause environmental degradation or are found to be harmful to human health.[56]

Principle 6 of the same document states that "the special situation and needs of developing countries, particularly the least developed and those most environmentally vulnerable, shall be given priority."[57]

One of the key decisions taken at the Rio Earth Summit was adoption of Agenda 21, called "a document of hope" because it attempts to address and overcome economic and ecological problems worldwide.[58] Agenda 21 was drafted to provide a comprehensive blueprint for the world in addressing many environmental issues, and the effects of toxic substances are one of those global issues. It does not presume to propose solutions that will solve completely all of the problems confronting humankind and the environment, but rather it introduces a series of actions by which local, regional, and international solutions can be identified and put into practice.

Agenda 21, Chapter 17 on the marine environment and Chapter 19 on "environmentally sound management of toxic chemicals, including prevention of illegal international traffic in toxic and dangerous products," contained the first international legal language regarding problems subsequently raised in the Stockholm Convention (*see* Chapter 4, below). This documentation of the problem of unrestricted use of chemicals and pesticides was an unprecedented benefit to the world as a check on the concept of human "progress," as meaning "economic and technological growth without regard to any impact on the biosphere."[59] As stated by a noted political scientist, we are now experiencing "an awakening of modern man to a new awareness of the human predicament on earth."[60]

Chapter 19 of Agenda 21 acknowledges the need for an international strategy for action on chemical safety. This call arose during the Preparatory Committee meetings leading up to the Rio Conference, and highlighted the importance of chemical management. The strategy was left undefined, however. At the end of the Rio Conference, this topic was adopted in the process that became the Intergovernmental Negotiating Committee for the Stockholm Convention. Chapter 19 of Agenda 21 sets forth a comprehensive set of rules regarding "Environmentally Sound Management of Toxic Chemicals Including Prevention of Illegal Inter-

---

56 *Rio Declaration on Environment and Development, at* http://www.unep.org/unep/rio.htm (last visited on Jan. 23, 2002).
57 *See id.,* at Principle 6.
58 *Id.*
59 UN Conference on Environment and Development, *Agenda 21, at* http://un.org/esa/sustdev/agenda21text.htm (last visited Jan 25, 2002).
60 L.K. CALDWELL, INTERNATIONAL ENVIRONMENTAL POLICY 9 (2 ed., Duke University Press 1990).

national Trade in Traffic in Toxic and Dangerous Products."[61] This approach was prompted by the increasing international concern that toxic products circulate in violation of existing national laws and cause harm far from the producing state.[62] Therefore, international cooperation in the prevention of illegal traffic of toxic substances, including POPs, is necessary.

The objective of chapter 19 of Agenda 21 is to strengthen international chemical risk assessment and to set forth guidelines for acceptable exposure levels for a number of chemical substances. This Chapter also helped to draw international attention to POPs by emphasizing the importance of dealing with all chemical substances that are "toxic, persistent and bioaccumulative and whose use cannot be adequately controlled."[63]

Six areas of activities identified in Chapter 19 of Agenda 21 established global strategy to promote chemical safety as follows:[64] a) expand and accelerate international assessment of chemical risks, b) establishment of a harmonized system for classification and labeling of chemicals, c) information exchange on toxic chemicals and chemical risks, d) establishment of risk reduction programs, e) strengthening of national capabilities and capabilities for management of chemicals, and f) prevention of illegal international trade in toxic and dangerous products.

Since UNCED in Rio de Janeiro, governments have recognized that environmental dangers do not respect national borders, no matter how developed the country is or how well guarded it is. Many governments, including that of Brazil, have reviewed their own domestic legislation in light of the results of UNCED and included new legislation modeled on Agenda 21 and the Rio Declaration. For example, a new law was adopted in Brazil regulating the proper disposal of containers of pesticides, which is the obligation of producers of these chemicals. Another example of a new Brazilian environmental law after UNCED is the 1998 law establishing environmental crimes, including pollution.

5. The UNEP Code of Ethics on the International Trade in Chemicals (1994)

The Code of Ethics on the International Trade in Chemicals is the outcome of a series of UNEP consultative meetings for private sector parties convened between May 1992 and April 1994 in accordance with UNEP Governing Council Decision 16/35 and Chapter 19 of Agenda 21.[65] Once drafted by UNEP, other entities endorsed the Code of Ethics. For example, during the first meeting of the Intergovernmental Forum on Chemical Safety (IFCS), in Stockholm, April

---

61 *Agenda 21*, ch. 19, at http://un.org/esa/sustdev/agenda21text.htm (last visited Apr. 5, 2002) (Environmentally Sound Management of Toxic Chemicals, Including Prevention of Illegal International Traffic in Toxic and Dangerous Products).

62 *See id.*, at art. 19.10. (United Nations General Assembly Resolutions 44/226 of December 22nd, 1989).

63 *See id.*, at ch. 19.49(b) & (c).

64 *See supra* note 246.

65 *UNEP Code of Ethics on the International Trade in Chemicals*, at http://irptc.unep.ch/ethics/english/CODEEN.html (last visited on Jan. 22, 2002).

1994, attended by technical and scientific experts appointed by member states of the United Nations, representatives agreed that the Code of Ethics should be applied widely by industries in all countries.

The Code of Ethics suggests disclosure of known chemical information about substances entering into international trade. This safety data sheet is more limited than the PIC procedure in that no explicit warnings of danger are included and, most significantly, no reply accepting the substance for import is required. In the event of import or export of certain chemical substances, the Code requires attachment to each shipment of a document, which supplies all known information about the substance. This document is known as the "safety data sheet," prepared by IFCS for all dangerous chemicals in international trade.[66] The safety data sheets constitute principles and guidance for the chemical industry and other private sector parties to enhance chemical safety in all countries.[67] At the second session of the Commission on Sustainable Development of the United Nations in New York, in May 1994, UNEP's previous recommendation to apply the voluntary Code of Ethics was endorsed within the United Nations as a whole.[68] This voluntary Code of Ethics was distributed by UNEP to 185 industries and business associations, 77 non-governmental organizations, and all governments and intergovernmental organizations around the globe, with an invitation to apply its provisions.[69]

The objective of the Code of Ethics is to set standards of conduct to encourage environmentally sound management of chemicals in international trade.[70] The Code of Ethics

> is a complement to the amended London Guidelines for the Exchange of Information on Chemical in International Trade, which address Governments and the scope of the Code is broader then that of the amended London Guidelines. By the implementation of this Code, the private sector parties are expected to enter into voluntary commitment to help achieve the objectives of the amended London Guidelines, i.e., to increase chemical safety and to enhance the sound management of chemicals in all countries through the exchange of information on chemicals in international trade.[71]

It was designed to be consistent with and complementary to existing instruments such as the UNEP London Guidelines and the FAO Code of Conduct, while avoiding duplication. Industries adopting the Code of Ethics committed to take measures to improve safety in production, management and international trade of chemicals. While some industries adopted both the Code of Ethics and other voluntary documents, such as FAO's Code of Conduct, other industries adopted

---

66 *Report on the Status of the Application of the Code of Ethics on the International Trade in Chemicals,* at http://chem.unep.ch/ethics/english/rep-en1.htm (last visited on Jan. 22, 2002).
67 *See id.*
68 *See supra* note 250.
69 *See supra* note 250.
70 *See id.* at Part I(1) (Objective).
71 *See id.,* at no. 3 (Introduction to the Code of Ethics).

none of them. Industries that adopt only the Code of Conduct are encouraged to make a declaration that their commitment is consistent with UNEP's Code of Ethics.[72] Unlike databanks such as the registry maintained by UNEP Chemicals (IRTPC), these voluntary codes attempted to raise consciousness among industries and consumers of pesticides about risks and recommended precautions through dissemination of information to users.

6. International Organization for the Sound Management of Chemicals (IOMC) (1995)

Another important body that was put together to assist governments worldwide in implementing Chapter 19 of Agenda 21 is the Inter-Organization Programme on the Sound Management of Chemicals (IOMC). The IOMC was established in 1995 as a mechanism for coordinating efforts of intergovernmental organizations in the field of chemical safety. It was created by several international organizations, including the United Nations Environment Programme (UNEP), International Labor Organization (ILO), Food and Agriculture Organization (FAO), World Health Organization (WHO), United Nations Institute for Training and Research (UNITAR), United Nations Industrial Development Organization (UNIDO), and the Organization for Economic Cooperation and Development (OECD), called "participating institutions."[73] The goals of the IOMC are to increase awareness and foster international cooperation and precaution in the field of chemical safety. Its main objective is to "promote coordination of policies and all activities by the participating institutions in order to observe the sound management of chemicals in relation to human health and the environment."[74] Chemical risk evaluation, harmonization of classification of chemicals, and information exchange on toxic chemicals and chemical risks are collected by IOMC.

---

72 *See id.*, at § IV(6).
73 UNEP: Established in 1972 as a program under ECOSOL and within UNEP, the chemical unit is the focus for all activities to ensure the global sound management of hazardous chemicals and to protect human health and the environment from impacts of toxic chemicals.
    ILO: This UN specialized agency with a tripartite constituency of governments, employers and employees has Chemical Safety as part of its mandate. It sets standards on workers exposure to chemicals (ILO # 170/1990) and the Prevention of Major Industrial Accidents (ILO # 174/1993).
    FAO: Assists member countries to improve their use and management of chemicals in agriculture.
    WHO: Its work in chemical safety is done largely through the IFCS, a joint program of WHO, ILO, and UNEP. It evaluates the risk to human health and the environment, methods for assessment, sound management of chemicals and risk reduction goals.
    UNIDO: The chemical industry branch was established in 1967, when the agency was created, but after the Bhopal accident in 1994 it became deeply involved in technical cooperation and occupational safety. Web site at http://www.unido.org.
    UNITAR: Promotes training and capacity building in chemical and waste management practices. Web Site at http://www.unitar.org.
    OECD: The chemical branch was established in 1978 to promote sound management of chemicals worldwide, integrate economic and chemical safety policies, and prevent unnecessary trade barriers. Web Site at http://www.oecd.org.
74 *UNEP Governing Council Decision 18/32* (May 25, 1995), at http://irptc.unep.ch/pops.

The work of chemical risk assessment and environmentally sound management of chemicals, used by IOMC in its coordination efforts, is actually performed by the Intergovernmental Forum on Chemical Safety ("IFCS").[75]

At the Rio Summit in 1992, the foundation was laid which helped build momentum towards the preparation of the Stockholm Convention; two years later, a cooperative entity called the IFCS was set up to further scientific understanding to underpin a new treaty. A number of steps were taken at regional and international levels that contributed to the development of the future international treaty on persistent organic pollutants. First, the Swedish government held a joint conference on April 25-29, 1994, in Stockholm with the International Labour Office (ILO), the United Nations Environment Programme (UNEP), and the World Health Organization (WHO). Called the "International Conference on Chemical Safety: For An Environmentally Sound Management of Chemicals,"[76] the Conference was in response to recommendations of Agenda 21, Chapter 19, paragraph 76. This meeting created the Intergovernmental Forum on Chemical Safety (IFCS).[77] International bodies were called upon to work together to strengthen their collaboration and extend it to other related international organizations, with the objective to make the IFCS "the nucleus for international cooperation on environmental sound management of toxic chemicals."[78] IFCS is a mechanism for cooperation among governments as they meet with intergovernmental and non-governmental organizations to more effectively integrate national and international efforts to promote chemical safety. It provides policy guidance and strategies that will enable governments to achieve harmonization in risk assessment and classification of chemical compounds. The World Health Organization serves as the administrative body and secretariat for the Forum.[79]

Under UNEP Governing Council Decision 18/32 concerning Persistent Organic Pollutants,[80] IOMC was invited to work jointly with the International Program on Chemical Safety (IPCS) and the Intergovernmental Forum on Chemical Safety (IFCS) to assess possible persistent organic pollutants, which would then constitute evidence for future action. This scientific study was conducted initially with twelve chemical compounds, analyzing their biochemistry, transport, and harmful effects on human health and the environment, and searching for economically viable substitutes to these substances. A final report was then produced entitled "A Review of the Persistent Organic Pollutants: DDT, Aldrin, Dieldrin,

---

75 The Web Site for the Intergovernmental Forum on Chemical Safety is at http://www.who.int/ifcs/.
76 Press Release, World Health Organization, WHO/36 (Apr. 22, 1994), at http://www.who.int/archives/inf-pr-1994/pr94-36.htm (last visited on February 19, 2002).
77 Res.1, IPCS/IFCS/94 (Apr. 29, 1994), available at http://www.who.int/ifcs/fs_res1.htm (last visited on Feb. 19, 2002) ("Resolution on the Establishment of the Intergovernmental Forum on Chemical Safety").
78 See id.
79 The Web Site for the Intergovernmental Forum on Chemical Safety is at http://www.who.int/ifcs/.
80 Id.

Endrin, Chlordane, Heptachlor, Hexachlorobenzene, Mirex, Toxaphene, Polychlorinated Biphenyls, Dioxins and Furans."[81] However, neither the IFCS nor the IOMC created any legal binding obligation on states.

This section has discussed seven documents or entities created by the international community to address the adverse effects of hazardous chemicals: pesticides, industrial chemicals, and byproducts of industrial activity. These statements, registers, procedures and documents are known as "soft law," a part of international environmental law without binding effect.

More binding measures to control POPs would require the creation of new international "hard law," either by recognizing customary international law or treaty law, that are binding on states.[82]

C. Customary Law and Principle 21

Customary law is acknowledged by a significant number of states as creating legal obligations, and accordingly is followed by states just the same as treaty law; indeed, customary law is one of the sources of international law along with treaty law, following Article 38 (1)(b) of the Statutes of the International Court of Justice.[83] In order to identify a principle or obligation as customary law, it is necessary to find both the elements of state practice and *opinion juris*, meaning the actual deeds and speeches of officials of a state invoking the principle, and evidence that they believe the principle creates a binding legal obligation for them.[84] Evidence of custom may be found in diplomatic correspondence, policy statements, press releases, speeches at international conferences by legal advisors or heads of state, executive decisions and practices, orders to armed forces or executive departments, etc. Duration, uniformity and consistency, and proof of local customs all may be considered in establishing the existence of custom, although there are no "bright line" rules for identifying customary law.[85]

Principle 21 of the Stockholm Declaration[86] is widely recognized as customary international law, and maybe also the precautionary principle, although scholars may debate the latter.[87] Principle 21 declares that: "all states have the responsibility to ensure that activities under their jurisdiction or control do not cause damage to the environment of other states or to areas beyond national jurisdic-

---

81 Full Report, IOMC, *at* http://www.chem.unep.ch/pops/indxhtms/asses0.html (Dec. 1995).
82 Statutes of the International Court of Justice, Article 38, on the sources of international law.
83 Henkin, Pugh, Schachter, Smit, INTERNATIONAL LAW: CASES AND MATERIALS, 3RD ED., West Publishing, St. Paul, Minn. 1993, pp. 54-57 *et seq.*
84 *See generally,* Ian Brownlie, PRINCIPLES OF PUBLIC INTERNATIONAL LAW, 4TH ED., Clarendon Press, Oxford, England, 1994, pp. 4-11.
85 *Id.*
86 United Nations, *Report of the United Nations Conference on the Human Environment*, document A/Conf.48/14/Rev.1, Chapter 1 (New York: 1972).
87 THE PRECAUTIONARY PRINCIPLE AND INTERNATIONAL LAW: THE CHALLENGE OF IMPLEMENTATION, 53-71 (David Freestone & Ellen Hey eds., 1996).

tion." This language is mirrored in the Preamble to the Stockholm Convention, which states:

> *Reaffirming* that States have, in accordance with the Charter of the United Nations and the principles of international law, the sovereign right to exploit their own resources pursuant to their own environmental and developmental policies, and the responsibility to ensure that activities within their jurisdiction or control do not cause damage to the environment of other States or of areas beyond the limits of national jurisdiction.

The reason why this principle is so important in the area of POPs is that some states which do not actually ratify the Stockholm Convention may still be under legal obligations by virtue of customary law not to cause harm to the territory of other states or to the shared commons by the release of POPs. One of the chemical characteristics of POPs, after all, is the long-range travel of the substances and to their persistence, meaning that these toxic substances do not degrade as they move in the atmosphere or in water. Uniquely, perhaps, in this subject it is unlikely that any state could be characterized as a "persistent objector," and therefore exempt from the customary law of POPs. The reason for this is that the science is clear regarding the risks and the causation of harm, and is not contested by any state, according to the discussions in the negotiating sessions leading up to the adoption of the text of the Stockholm Convention. In addition, no state has declared that it believes itself free to disregard the risks and continue indefinitely in the manufacture, transport, and use of POPs to the degree that might constitute their rejection of customary law. The single exception, of course, is those states who have requested additional time in which to comply with what they recognize as legal obligations to reduce or eliminate the use of POPs, as listed in the exemptions filed with the Secretariat at the time of signing the Stockholm Convention. These exemptions are valid for five years, subject to a single renewal request. Since this is a matter of reservations to the treaty, it could by extension be construed as an objection to being legally bound by any customary law requiring the reduction or elimination of the release of POPs. Those states claiming the exemption are doing so in the name of public health and for a temporary period only until they can control communicable diseases such as malaria or find alternative substances that are economically viable. This distinction appears to be different from the traditional "persistent objector" status, whereby a state declares itself free from any restriction or legal constraint which otherwise might obligate it under a principle of customary law; in this case, during the limited period of the exemption only, a state might defend itself against claims of violation of customary law. However, once the exemption period expires, that state has declared itself ready to be bound by legal obligations.

Applying the traditional analysis of customary law to POPs, we can find evidence of state practice and *opinion juris* in the diplomatic correspondence and speeches of states at international conferences and the negotiating sessions for the Stockholm Convention, where many leaders denounced the hazards caused to humans and the environment from POPs. There is municipal legislation from selected states which incorporates the legal obligations on a domestic level as an indication of their recognition as binding law on the international level by those states. In addition, there are some court cases, such as one pending in Brazil's Federal Court in the state of Rio Grande do Sul addressing a chemical company's liability for harm from site contamination and harm to human health due to POPs. The adoption of the language in the Stockholm Convention is also evidence of the customary law relating to POPs.

In addition, states which have signed the Stockholm Convention but not yet ratified it are, under provisions of the Vienna Convention on the Law of Treaties, expected to observe their international treaty obligations in the interim while pursuing ratification domestically. Accordingly, even absent the entry into force of the Stockholm Convention, we can see that obligations regarding POPs and prevention of harm from their release, do exist in customary law and prior to ratification of the treaty itself.

In conclusion, then, oftentimes customary law principles create legal obligations and crystallize into treaties, which are hard law. In the case of POPs, it appears that the development of international law has moved from domestic law and regulation in some states, to international soft law in United Nations declarations and UNEP/FAO/WHO documents, to customary law in the sense of the applicability of Principle 21 and perhaps the precautionary principle to the transboundary movement and release of POPs, and directly to treaty law in the new Stockholm Convention.

D. "Hard Law" Measures to Regulate Hazardous Chemical Substances—The First Multilateral Treaties

Following the development of domestic, binational, and regional law and international soft law on POPs, the international community turned to the challenge of drafting international treaties regulating hazardous chemicals. Several preliminary treaties discussed below are: the Basel Convention on the Control of Transboundary Movements of Hazardous Wastes and Their Disposal; ILO's Chemicals Convention and Recommendation; the Bamako Convention on the Ban of the Import into Africa and the Control of Transboundary Movement of Hazardous Wastes within Africa; the Convention on Long-Range Transboundary Air Pollution (LRTAP); the Convention on the Prior Informed Consent Procedures for Certain Hazardous Chemicals and Pesticides in International Trade (the Rotterdam Convention).

1. Basel Convention on the Control of Transboundary Movements of Hazardous Wastes and Their Disposal (1989)

Environmental concerns arising from disposal of hazardous substances in developing countries first gained international attention in the late 1980s, when many unfortunate disasters occurred from dumping toxic wastes in Africa. For example, in 1987, two brokers from the waste sector firms Ecomar and Jelly Wax, Gianfranco Raffaeli and Renato Pent respectively, signed an agreement with

Nigeria businessman Sunday Nana to use his estate for the storage of 18,000 drums of highly toxic and radioactive waste, including 150 tons of PCBs. The waste was brought into the country as "residual and allied chemicals" and entered through the port of Koko.[88] In 1988, Guinea-Bissau, a country with a Gross National Product of US$150 million, was offered a contract worth US$600 million dollars to allow 15 tons of toxic wastes over five years to be imported into their country from European and American waste brokers.[89] In 1987, Lebanon's right wing Christian militia was offered a large sum of money to import 15,000 barrels of toxic waste and 20 containers from Italy. Since they had no idea what to do with the waste, most of it was simply dumped across the countryside, poisoning soil and water supply.

The reason for sending hazardous waste to developing countries is primarily economic. First, with the growth in awareness among industrialized countries of the dangers posed by unsound disposal of hazardous waste, stricter national legislation demanded costly measures for environmental protection and sound chemical management. Developing countries did not, and some still do not have, national legislation prohibiting such pernicious practices. The average cost of waste disposal in Africa is between US$2.50 and US$50 per ton, while in most industrialized countries this cost is between US$100 and US$2,000 per ton.[90]

With that scenario in mind, UNEP started to push for work on an international legally binding agreement to control trade of toxic wastes. In 1987, delegates started negotiating a treaty to put the onus on exporting countries to ensure that hazardous wastes are managed in an environmentally sound manner within the country of import. The Basel Convention on the Control of Transboundary Movement of Hazardous Wastes and Their Disposal[91] was signed, after eighteen months of negotiations, by thirty-five of the 116 countries participating in the Conference of the Plenipotentiaries on March 22nd, 1989 in Basel, Switzerland. The Convention entered into force on May 5th, 1992. As of January 2nd, 2002, 149 Parties have ratified the Basel Convention.[92] It was not an easy negotiation due to differing priorities of industrialized and developing countries. One of the most difficult Articles to negotiate was the definition of "wastes," and "hazardous wastes."[93] Second, developing countries wanted to see a total ban on all transboundary movements of hazardous substances, but industrialized nations

---

88 K. Arti, *Toxic Trade with Africa*, 1989 ENVT., SCI. & TECH. J. 24.

89 Jean-Paul Dufour & Corine Denis, *The North's Garbage Goes South*, 32 WORLD PRESS REVIEW 30-35 (Nov. 1988).

90 KATHATINA KUMMER, INTERNATIONAL MANAGEMENT OF HAZARDOUS WASTES: THE BASEL CONVENTION AND RELATED LEGAL RULES 10 (Oxford University Press. 1995).

91 *Basel Convention on the Control of Transboundary Movement of Hazardous Wastes and Their Disposal*, UNEP Doc. I.G. 80.3, *available at* http://www.basel.int/text/con-e.htm (last visited on Feb. 4, 2002).

92 *See id. See also Status of Ratification of the Basel Convention, at* http://www.basel.int/ratif/ratif.html (last visited on Feb. 18, 2002).

93 *Summary of the Fifth Conference of the Parties to the Basel Convention on the Control of Transboundary Movements of Hazardous Wastes and their Disposal, 1999* EARTH NEGOTIATIONS BULLETIN 20, *available at* http://www.iisd.ca/linkages/download/pdf/enb2006e.pdf.

fiercely rejected this proposal. The United States did not want to see different regulations governing foreign and domestic disposal.[94] It is felt that the greater the discrepancy between domestic and international regimes, the more difficult it is for nations to ratify an international treaty. Countries would have to change their national legislation in order to comply with the treaty, and this can be a time consuming and, in many cases, a very political battle. Protection of each country's sovereignty was another important issue because developing nations refused to accept an international inspector checking their landfill operations. Since most developing countries do not have the expertise to manage hazardous waste or the technology to do so, they demanded transfer of technology from developed states to help them comply with the provisions of the treaty, as well as additional financial assistance.[95]

The Basel Convention[96] delineated two major objectives to be accomplished: a) environmentally sound management and disposal of hazardous wastes, and b) control of the transboundary movement of these substances. The Convention defines "wastes" as substances or objects, which are "disposed of or are intended to be disposed or are required to be disposed of by provisions of national law."[97] All chemical substances identified as POPs when defined as wastes and subject to transboundary movement will fall into the category of hazardous waste under the Basel Convention. Annex I of the Basel Convention specifically identifies several wastes that are persistent organic pollutants, such as PCBs, dioxins, furans, and all substances or wastes which, if released, present or may present immediate or delayed adverse impacts to the environment by means of bioaccumulation and/or toxic effects upon biotic systems.[98]

The scope of the Basel Convention covers all substances or objects defined as "waste;" this waste must fall into the definition of "hazardous"[99] and it must be subject to transboundary movement.[100] The Basel Convention can thus be applied to POPs in the event of their disposal. POPs would fall into the definition of "hazardous waste" only if disposed of and not otherwise.

2. International Labour Organization's Chemicals Convention and Recommendation (1990)

One other treaty that may be related to POPs is the Chemicals Convention and Recommendation from the International Labor Organization. The emphasis here was on workers' rights and protection from exposure to certain chemical substances that had carcinogenic, toxic or allergenic effects in the work environment.

---

94   51 Fed. Reg. 28670 (1986).
95   *See infra*, Article 10 (4).
96   *Basel Convention on the Control of Transboundary Movements of Hazardous Wastes and their Disposal, at* http://www.basel.int/text/con-e.htm (last visited on Jan. 29, 2002).
97   *See id.* at art. 2 (Definitions).
98   *See id.* Annex I (Categories of Wastes to be controlled).
99   *See id.* art. 1(1)(a).
100 *See id.* art. 1.

The treaty was not universal, as it was binding only on the members of the International Labor Organization, a United Nations specialized agency founded in 1919 to promote social justice and international harmonization of human and labor rights. Along with the state members, unions and non-governmental organizational members help to formulate international labor standards in the form of conventions and recommendations establishing minimum standards of basic labor rights, among which are issues affecting occupational safety and health at work.[101] The Governing Body of the International Labor Organization, at its 77th Session on June 6th, 1990, adopted Chemical Convention No 170[102] and a series of recommendations (No. 177)[103] on safety in the use of chemicals at work. The main objective of this Convention was to inform workers about the harmful effects of toxic chemicals and enhance their protection at work, incidentally protecting the environment. The Convention establishes comprehensive criteria for classification of chemical substances, including their toxic properties, carcinogenic effects, allergenic and sensitizing effects, and teratogenic, mutagenic and reproductive system effects.[104] It provides workers with information about chemicals in general use at the workplace, and about appropriate preventive measures to be used for the production, handling, storage, transport, disposal and treatment of chemicals and measures for maintenance, repair and cleaning of equipment or containers for chemicals.[105] One of the key aspects of this Convention is the labeling requirement for marking hazardous chemicals in such a way that workers can easily understand and recognize them and take precautions while using them in any way.[106]

3. The Bamako Convention on the Ban of the Import into Africa and the Control of Transboundary Movement of Hazardous Wastes within Africa (1991)

After the adoption of the Basel Convention in March 1989, the Organization of African Unity (OAU) expressed concern that the Basel treaty did not adequately address the issues of transboundary movement of hazardous waste by developing nations. Perhaps the major concern of the OAU was that the Basel Convention did not completely ban all transboundary movements of hazardous wastes. This prompted African states to draft and adopt a regional agreement called the Bamako Convention,[107] which establishes a total ban on the import of all

---

101   ILO Mandate at http://www.ilo.org/public/english/about/mandate.htm.
102   *ILO Chemicals Convention No. 170, at* http://www.ilo.org/public/english/sitemap.htm (last visited on Jan. 18, 2002) (full text).
103   *ILO Chemicals Recommendation No. 177, at* http://www.ilo.org/public/english/sitemap.htm (last visited on Jan. 18, 2002) (full text).
104   *Id.* § 6(a)-(g).
105   *See ILO Chemicals Convention No. 170* art. 2(c)(i)-(vii), at http://www.ilo.org/public/english/sitemap.htm (last visited on Jan. 18, 2002).
106   *See id.* art. 7(2) (Labeling and Marking).
107   *The Bamako Convention on the Ban of The Import Into Africa and the Control of Transboundary Movement of Hazardous Wastes within Africa, at* http://www.londonconvention.org/Bamako.htm (last visited on Feb. 4, 2002).

hazardous wastes into Africa and limits the transfer of wastes within Africa. Negotiators decided to distinguish between waste generated inside and outside Africa, prohibiting all imports of foreign hazardous wastes into Africa but not those generated within the region. The major criticism of the Basel Convention by the OAU was that it merely controlled the movement of waste, and that was insufficient to protect people in Africa.[108] OAU countries agreed that despite the fact that transfers of hazardous waste provide economic benefit on one hand, lack of financial resources, technology or expertise to manage their disposal facilities in an environmentally sound manner pose a great threat to human health and the environment. The Bamako Convention establishes four main objectives: a) a ban on the import of hazardous waste into Africa; b) the liability of waste generators for damage caused by their waste; c) assistance from developed countries in monitoring and controlling hazardous waste movements; and d) the use of ecologically sound methods of disposal of hazardous wastes.[109] Similar to the Basel Convention technique of listing all targeted substances and materials to be regulated in one separate annex, the Bamako Convention lists all categories of wastes which are hazardous and subject to the provisions of the Convention in Annex I. Therefore, if a substance is listed in this annex, according to Article 2 (1)(a) of the Bamako Convention, it is hazardous. Among these substances are a number of persistent organic pollutants such as halogenated organic solvents, dioxins and furans.[110] Annex II of the Convention also includes all POPs under the category "Ecotoxic," meaning "any substance or waste which, if released, present or may present immediate or delayed adverse impacts to the environment by means of bioaccumulation and/or toxic effects upon biotic systems."[111]

4. The Convention on Long-Range Transboundary Air Pollution (LRTAP) and its Protocol on Persistent Organic Pollutants (1998)

Since the risk posed by Persistent Organic Pollutants was becoming a major concern to many countries, scattered actions to protect human health and the environment began to appear. At a regional level of action, the United Nations Economic Commission for Europe drafted a treaty to deal with problems of air

---

108 African Countries party to the Bamako Convention signed in January 1991 by the Conference of Environment Ministers, OAU, at Bamako, Mali: Algeria, Angola, Benin, Botswana, Burrina Faso, Burundi, Cameroon, Cape Verde, Central African Republic, Chad, Comoros, Congo, Cote D'Ivoire, Djibouti, Egypt, Equatorial Guinea, Ethiopia, Gabon, Gambia, Ghana, Guinea, Guinea Bissau, Kenya, Lesotho, Liberia, Socialist People's Libyan Arab Jamahiriya, Madagascar, Malawi, Mali, Mauritania, Mauritious, Mozambique, Namibia, Niger, Nigeria, Rwanda, Sahrawi Arab Democratic Republic, Sao Tome and Principe, Senegal, Seychelles, Sierra Leone, Somalia, Sudan, Swaziland, Tanzania, Togo, Tunisia, Uganda, Zaire, Zambia, and Zimbabwe.

109 J. Wylie Donald, *The Bamako Convention as a Solution to the Problem of Hazardous Waste Exports to Less Developed Countries*, 17 COLUM. J. ENVTL. L. 419, 429-30 (1992).

110 *See The Bamako Convention on the Ban of The Import Into Africa and the Control of Transboundary Movement of Hazardous Wastes within Africa* Annex I, *at* http://www.londonconvention.org/Bamako.htm (last visited on Feb. 4, 2002).

111 *See id.* Annex II(H12) (Ecotoxic).

pollution in the transboundary context.[112] The resulting Convention on Long-Range Transboundary Air Pollution (LRTAP)[113] is considered to be the first legally binding international instrument to address air pollution on a broad regional basis. Once scientific evidence demonstrated the effects of sulfur emissions in continental Europe and acidification of bodies of fresh water in Scandinavia,[114] it was soon understood that air pollutants can be transported thousand of miles before they are precipitated to soil and lakes. There is also evidence that sulfur oxides and acid rain damages crops, delays forest growth, destroys surfaces of stone buildings and monuments, and contaminates drinking water.[115] European countries such as Norway and Sweden suffered from acid soil contamination caused by long-range transport of air pollutants. The transboundary aspect of this problem prompted these two countries to request a European solution from the United Nations Economic Commission for Europe.[116] Around the same time, the Global Environmental Monitoring System (GEMS), a program of the United Nations Environment Programme, created sulfur dioxide monitoring stations to collect data on the flow of sulfur dioxide across national borders.[117] There were monitoring stations in Norway, Sweden, Finland, Hungary, Belgium, Ireland, Austria, Poland, East and West Germany, the Netherlands and Italy.[118] For this particular purpose, GEMS created the European Monitoring and Evaluation Program (EMEP), funded by UNEP and coordinated by the World Meteorological Organization.[119]

Shortly thereafter, the Convention was opened for signature on November 13th, 1979, and the treaty entered into force in 1983. As of February 19, 2002, the LRTAP Convention has 48 Parties to the treaty, and 8 Parties to the POPs Protocol.[120]

Perhaps the most important contribution of this Convention was to serve as the first step in stimulating action on a regional level to combat an international environmental harm, pollution of the air. The most important aspect of this con-

---

112 UNEP, *Final Report of the Meeting of the Intergovernmental Forum on Chemical Safety Ad Hoc Working Group on Persistent Organic Pollutants*, (Manila, Philippines) June 21-22, 1996, at http://irptc.unep.ch/pops/indxhtms/manwgrp.html (last visited on Jan. 25, 2002).
113 *The Convention on Long-Range Transboundary Air Pollution* (LRTAP), (Nov. 13, 1979), at http://www.unece.org/env/lrtap/conv/lrtap_c.htm.
114 AMAP, Assessment Report: Arctic Pollution Issues 860 (1998).
115 *See* Bureau of Oceans and International Environment and Scientific Affairs, US Department of State, Second Report on the US-Canada Research Consultation Group on the Long Range Transport of Air Pollutants. (Nov. 1980).
116 Telephone interview with Mr. Henning Wuester, UN/ECE Environment and Human Settlements Division (Jan. 18, 2002).
117 *See Cooperative Program for Monitoring and Evaluation of the Long Range Transmission of Air Pollutants in Europe*, UN Doc. ECE/ENV 15, Annex II (1977).
118 *See* LRTAP, *supra* note 298.
119 *See* LRTAP, *supra* note 298.
120 Telephone Interview with Keith Bull, Secretary to the Executive Body to the Convention on Long-range Transboundary Air Pollution, (Feb. 19, 2002).

vention for POPs was one of the subsequently adopted Protocols that specifically addressed the issue of persistent organic pollutants.[121]

In 1989, the Executive Body for the Convention asked a group of experts from various fields of science, such as chemistry and biology, for a full report on Persistent Organic Pollutants.[122] As a follow-up, the experts met four times between 1991 and 1994, producing a substantial report assessing the state of scientific knowledge on POPs emissions, transport, and impacts, which served as the foundation for the draft of a negotiating text for the POPs Protocol to LRTAP. The Task Force of Experts on Persistent Organic Pollutants' report was on the persistence, toxicity, biomagnification, and effects of long-range transport of fifteen chemical substances: aldrin, dieldrin, endrin, chlordane, chlordecone, DDT, heptachlor, hexabromobiphenyl, hexachlorobenzene, mirex, PAHs, PCBs, dioxins, furans, and toxaphene.[123] Based on this report, the United Nations Economic Commission for Europe (UN ECE)[124] held a ministerial meeting in Åarhus, Denmark, from June 23-25, 1998, and adopted the Protocol to Regulate Long Range Air Pollution by Persistent Organic Pollutants.[125]

The effort to address POPs in the context of the LRTAP Convention came principally from Canada and Sweden. Sweden, always active in compliance with the LRTAP Convention, was greatly affected by POPs due to the tendency of these substances to accumulate in the Arctic. Canada, another state also suffering the effects of POPs in the Arctic, responded to pressure from their indigenous populations, especially the Inuits in the north of Canada. There was solid science proving the adverse effects on these Arctic people because ocean and air currents carry POPs northward, where they are precipitated in great quantities, contaminating mammals in the food chain for humans in the Arctic. Sweden wanted to include a larger list of substances as POPs, at one point offering over one hundred substances to be included. The United States pushed strongly to limit the list, keeping it to only 16 chemical compounds. The original idea by Sweden and Canada, the two leading negotiators of the LTRAP Convention, was to make a framework convention under which a series of protocols would be incorporated, each addressing a single substance, such as sulfur dioxide.

Since LRTAP was a regional instrument, Canada and Sweden next pushed for international action on POPs, an initiative that resulted in the Stockholm Conven-

---

121 *LRTAP Protocol on Persistent Organic Pollutants (POPs)* (June 24, 1998), at http://www.unece.org/env/lrtap/protocol/98pop.htm (last visited on Jan. 18, 2002).
122 J.M. Pacyna, E. Voldner, G.J.Keeler, & G. Evans, *Proceedings of the first workshop on emissions and modeling of atmospheric transport of persistent organic pollutants and heavy metals* (May 6-7, 1993), at http://www.nilu.no/projects/ccc/reportlist.html#workshop.
123 *See* Report, Mr. Lars Nordberg, Executive Secretary of the Economic Commission for Europe, *at* http://www.chem.unep.ch/pops/POPs_Inc/proceedings/stpetbrg/nordberg.htm (last visited on Jan. 18, 2002).
124 UN/ECE–United Nations Economic Commission for Europe region covers the Russian Federation, the Newly Independent States, Central and Eastern Europe, Western Europe, Canada, and the United States.
125 *See* LRTAP POPs PROTOCOL, *supra* note 306.

tion. The information collected in the process of negotiating the POPs Protocol under the LRTAP Convention helped to publicize the issues and facilitate the negotiation of the POPs Stockholm Convention. As a Swedish official noted, when UNEP began negotiating the Stockholm Convention, they

> benefited from previous successful work within LRTAP, by being able to partly build on an already existing shared value base regarding issues of transboundary air pollution.[126]

Sufficient scientific data was gathered in the course of preparing the POPs Protocol to LRTAP Convention to serve as evidence of the effect of POPs and to further develop a general consensus for future global action on POPs. For example, the list of chemical substances identified by UNEP and included in Governing Council Decision 18/32[127] originated from the work carried out for the preparation of the Protocol. The Intergovernmental Negotiation Committee for the Stockholm Convention on persistent organic pollutants used much of the data used to strengthen the work carried out under the LRTAP Convention and Protocol, which also laid the ground for developing general principles of international cooperation in the field of transboundary pollution.

The UN ECE serves as the Secretariat. Sweden, Norway and Finland, the first to identify the threat to human populations from these hazardous chemicals, urged European states to develop the protocol on POPs for their own region. This was the first negotiated regional agreement on persistent organic pollutants, intended to serve as a model for a global instrument to eliminate POPs.[128] When the negotiations began on the Stockholm Convention, the Economic Commission for Europe provided technical expertise to UNEP, the lead agency for the global treaty.

The objective established in this Protocol is to "control, reduce, or eliminate discharges, emissions, and losses of Persistent Organic Pollutants into the environment."[129] Article 3 of the POPs Protocol to the LRTAP Convention contains mandatory measures to eliminate and limit production and use of persistent organic pollutants.[130] Sixteen chemical compounds, all of them POPs, were divided into three annexes to the POPs Protocol to LRTAP. Annex I lists all chemical substances intended for immediate elimination (aldrin, chlordane, chlordecone, DDT, dieldrin, endrin, heptachlor, hexabromobiphenyl, mirex, PCB, and toxa-

---

126 HENRIK SELIN, PROTECTING INTERNATIONAL COMMONS: BRIEF COMMENTS ON CURRENT ATTEMPTS AT ESTABLISHING NEW INTERNATIONAL INSTITUTIONS ON PERSISTENT ORGANIC POLLUTANTS (POPs) (1997).
127 UNEP, *Governing Council 9th Meeting of May 25th 1995*, Decision 18/32, *at* http://irptc.unep.ch/pops (last visited on Jan. 24, 2002).
128 *See* Daniel Pruzin, *Air Pollution: UN ECE Draft Protocol Concluded on Heavy Metals, Persistent Organics*, 21 Int'l Env't Rep. (BNA) No. 4, at 141 (Feb. 18, 1998).
129 *See* LRTAP POPs PROTOCOL, *supra* note 306, art. 2 (Objective).
130 *See id.* art. 3 (Basic Obligations).

phene).[131] Annex II lists those substances needing restrictions in use, such as DDT, HCH,[132] and PCB,[133] rather than a total ban. Annex III requires application of best available technology[134] to limit air emissions of dioxins, furans, hexachlorobenzene, and polycyclic aromatic hydrocarbons (PHAs).[135] These chemical emissions are unintended substances released into the atmosphere as a by-product of waste incineration, combustion, and metal production.[136] The Stockholm Convention applies 12 of the 16 chemicals addressed on the POPs Protocol to the LRTAP Convention, representing those chemicals on which global consensus existed regarding their persistence, toxicity, and bioaccumulation.

5. The Convention on the Prior Informed Consent Procedures for Certain Hazardous Chemicals and Pesticides in International Trade (The Rotterdam Convention, 1998)

The voluntary nature of the Prior Informed Consent procedures adopted under Agenda 21, Article 19 and the amended UNEP London Guidelines and FAO Code of Conduct has been noted above. The failures of the existing PIC system to develop an infrastructure and information systems on chemical management, many have been due to the following facts:

a) Many partners were failing to send notifications of the chemicals banned or restricted;[137]

b) Because it was a voluntary measure, there were no mechanisms to enforce compliance;

c) Most of the language of the Code and Guidelines were vague and subject to broad interpretation;[138]

d) Following the PIC procedures, industry would incur additional cost, and therefore incentives for illicit trade were created;[139]

---

131 *See id.* Annex I.
132 Hexachlorocyclohexane—HCH (CAS: 608-73-1) is a chemical that appears in the market with trade name Lindane, Agronexit, Lindafor, Gamma BHC, Kwell (shampoo). It is used for seed treatment, fumigant against flies and cockroaches. JOHN HARTE, CHERYL HOLDREN, RICHARD SCHNEIDER, AND CHRISTINE SHIRLEY, TOXICS A TO Z: A GUIDE TO EVERYDAY POLLUTION HAZARDS (University of California Press 1991).
133 *See LRTAP POPs PROTOCOL, supra* 306, Annex II.
134 *See id. LRTAP POPs PROTOCOL*, Annex V (Best Available Techniques to Control Emissions of Persistent Organic Pollutants from Major Stationary Sources), *at* http://www.unece.org/env/lrtap/protocol/98pop_a/annex5.htm (last visited on Jan. 18, 2002.
135 *See id.* Annex III.
136 B.D. Eitzer & R.A. Hites, *Atmospheric Transport and Deposition of dioxins and furans*, 23 Envtl. Sci. Tech., No. 11, at 1396-1401.
137 Janet Raloff, *PIC and Choose: A Toxic-Imports Accord*, Science News (Sept. 19, 1998).
138 *See id.*, at 582.
139 Michael P. Walls, *Chemical Exports and the Age of Consent: The High Cost of International Export Control Proposals*, 20 N.Y.U. J. Int'l L. & Pol. 753-766 (1988).

Chapter 3

e) The was no financial mechanism to help out developing nations to implement PIC measures;[140]

These factors led UNEP to initiate discussions on a legally binding instrument to regulate international trade in chemicals.[141]

In November 1994, at the 107th meeting of the FAO Council,[142] a decision was made to prepare a draft of a PIC Convention under the FAO/UNEP Joint Program on the implementation of PIC, based in part on the amended FAO Code of Conduct.[143] UNEP's Governing Council also concurred with the FAO initiative, and on May 26th, 1995, UNEP Governing Council adopted UNEP Decision 18/12, creating an Intergovernmental Negotiating Committee (INC)[144] to draft an international legal instrument detailing PIC procedures. The decision:

> authorized the Executive Director of UNEP to prepare for and convene, together with the Food and Agriculture Organization of the United Nations and in consultation with Governments and other relevant international organizations, within available resources, an intergovernmental negotiating committee, with a mandate to prepare *an international legally binding instrument* for the application of the prior informed consent procedure for certain hazardous chemicals in international trade.[145]

The first session of the Intergovernmental Negotiating Committee (INC) was held from 11-15 March 1996 in Brussels with delegates from more then 80 countries, the European Commission, a number of specialized agencies, and NGOs. Between March 1996 and March 1998, the INC met five times and completed a draft of the PIC Convention,[146] which was presented for adoption at the Diplomatic Conference in Rotterdam, September 10-11, 1998. More than 100 countries, the European Commission, specialized agencies, intergovernmental and non-governmental organizations attended the Rotterdam Conference for the approval of the PIC Convention.[147] As from January 15th, 2002, seventy-three countries and the European Community have signed and eighteen countries have ratified the PIC Convention under UNEP and FAO

---

140 *See* Raloff, *supra* note 322, at 197.
141 Report, International Institute for Sustainable Development, *A Brief Introduction to Chemical Management*, at http://www.iisd.ca/chemical/chemicalsintro.html (last visited on Jan. 24, 2002).
142 FAO Council 107th Meeting Decision CL 107/11, *Progress towards the development of legally binding instruments concerning the Prior Informed Consent Procedure* (Jan.28, 2002) (fax from FAO Headquarters to the author).
143 *See A Brief Introduction to Chemical Management, at* http://www.chem.unep.ch/pic/h5.html. The FAO/UNEP joint program was established to provide guidance to the Joint Secretariat on the development and implementation of the PIC procedure.
144 UNEP Governing Council Decision 18/12, *at* http://www.chem.unep.ch/pic/h5.html (last visited on Jan. 23, 2002).
145 *See id.* ¶ 1.
146 UNEP, *Convention on the Prior Informed Consent Procedure for Certain Hazardous Chemicals and Pesticides in International Trade, at* http://www.pic.int/finale.htm (last visited on Jan. 24, 2002).
147 *See id.*

auspices.[148] The PIC Convention does not regulate either the use or the production of hazardous chemicals, but rather their international movement as import or export. Through the treaty's provisions, prior informed consent procedures that once operated on a voluntary basis are now made legally binding for all States party to the Convention. The Convention will enhance information exchange and informed consent in international trade in chemicals and pesticides. During the process of negotiating the Convention, countries such as Malaysia, then speaking for the G-77, not only demanded that PIC procedures be made mandatory, but also demanded a ban on the export of domestically prohibited chemicals from countries that are members of the Organization for Economic Cooperation and Development (OECD) to other countries.[149] Taking the lead in determining the scope of the Convention, European countries represented by Belgium wanted to include measures to prohibit production of hazardous chemicals and to adopt phase-out procedures.[150] This would be done through the negotiation of future protocols on chemicals, creating a dynamic legal framework. NGOs applauded this initiative; however, Canada, the United States, Australia, and other non-European OECD countries vigorously opposed it on the grounds of cost.[151] At the UNEP experts' meeting in 1996 in Manila, Belgium and the Netherlands suggested phasing out POPs and other additional measures on chemical substances and the drafting of a "framework chemicals convention;" this proposal was opposed by the United States and Australia. Having the European countries in favor of a broader legal instrument to regulate chemical substances on one side, and strong opposition from the United States, Canada, and Australia on the other, developing countries trying to choose a position afraid of not being able to comply with a new treaty if the scope were broad.[152] The UNEP Governing Council decided to limit the negotiations in 1996 to the PIC procedures and later have a separate negotiation on persistent organic pollutants.[153] This enabled negotiations of the PIC Convention to begin and be concluded in 1998.

One of the key issues in the PIC Convention is the inclusion of chemical substances on the PIC list. The criteria for this inclusion was established by reaching a consensus between the threats posed by these chemicals to human health and the environment, and sound scientific evidence to back up

---

148 *Status of Signature and Ratification of the PIC Convention, at* http://www.pic.int/finale.htm (last visited on Jan. 24, 2002).
149 UNEP Governing Council, 1st Sess., UNEP Doc. UNEP/FAO/PIC/INC.1.3 (1996), *at* http://www.chem.unep.ch/pic/h3.html (last visited on Feb. 4, 2002).
150 *See id.*
151 *See* International Institute for Sustainable Development, *supra* note 326, at § 20 (Views on prohibition of use or the phasing out of hazardous chemicals).
152 *See id.*
153 UNEP Governing Council, 4th Sess., UNEP Doc.UNEP/FAO/PIC/INC.4/2 (1997), *at* http://www.chem.unep.ch/pic/incs/4/english/h3.html (last visited on Feb. 4, 2002).

the decision.[154] New chemicals may be added by the Conference of the Parties to the 27 chemicals initially included, divided into the following categories:[155]

a) Pesticides: 2.4.5-T, aldrin, captafol, chlorobenzilate, chlordane, chlordimeform, DDT, dieldrin, dinoseb, 1,2-dibromoethane (EDB), fluoroacetamide, HCH, heptachlor, hexachlorobenzene, lindane, mercury compounds, pentachlorophenol and certain formulations of methyl-parathion, methamidophos, monocrotophos, parathion, phosphamidon.

b) Industrial Chemicals: crocidolite, polybrominated biphenyls (PBB), polychlorinated biphenyls (PCB), polychlorinated terphenyls (PCT), tris(2,3 dibromopropyl) phosphate.

Other chemical substances may be added to the PIC Convention upon recommendation from the Chemical Review Committee, which will be comprised of government-designated experts on a basis of equitable geographical distribution.[156]

As mentioned above, the Convention aims to promote shared responsibility between exporting and importing countries for protecting human health and the environment against the harmful effects of hazardous chemicals.[157] The provisions concerning information exchange include:[158]

a) The requirement for a Party to inform other Parties of each ban or severe restriction on a chemical it implements nationally;

b) The possibility for a developing country Party or a Party with an economy in transition to inform other Parties that it is experiencing problems caused by a severely hazardous pesticide used in its territory;

c) The requirement for a Party that plans to export a chemical that is banned or severely restricted for use within its territory to inform the importing Party that such export will take place before the first shipment and annually thereafter;

d) The requirement that an exporting Party, when exporting chemicals that are to be used for occupational purposes, shall ensure that a safety data sheet that follows an internationally recognized format, setting out the most up-to-date information available, be sent to the importer, and

e) The requirement that exports of chemicals included in the PIC procedure and other chemicals that are banned or severely restricted domestically, when exported, are subject to labeling requirements that ensure adequate availability of information with regard to risks and/or hazards to human health or the environment.

---

154   Telephone Interview with Dr. Sergia Oliveira de Souza, Department for Environmental Quality, Brazilian Ministry of the Environment, (Jan. 12, 2002).
155   PIC Convention, Annex III, *at* http://www.pric.int/pic (Chemicals subject to the PIC Procedure).
156   *See id.*, art. 18(6).
157   *See id.*, art.1 (Objectives).
158   *See id.,* art. 14 (Information Exchange).

The PIC Convention cannot be seen as the ultimate solution for the regulation of the international trade in hazardous chemical substances, but it is a step towards a more extensive regulatory scheme. At the end of the Diplomatic Conference of the PIC Convention, UNEP Executive Director Dr. Klaus Töpfer noted that

> building sustainability into the chemical industry and the sectors of society that process, use, trade, and dispose of chemicals is the ultimate challenge for the next millennium and underscores the importance of collaboration between the chemical industry and the environmental community.[159]

Cooperation is indeed vital to solve the problem, and the PIC procedure has provoked international debate over problems of chemicals used in developing countries. However, one cannot only employ efforts to act in the sphere of prevention of trade in hazardous chemicals. Rather there needs to be an equal emphasis on preventing their production and eliminating the existing stockpiles of chemicals around the world.

---

159 *Conference on Plenipotentiaries on the Convention on the Prior Informed Consent Procedure for Certain Hazardous Chemicals and Pesticides in International Trade*, Earth Negotiations Bull. (Sept. 14, 1998), *available at* http://www.iisd.ca/linkages/download/asc/enb1511e.txt.

CHAPTER 4

THE FIRST GLOBAL TREATY ON PERSISTENT ORGANIC POLLUTANTS (POPs): NEGOTIATING THE STOCKHOLM CONVENTION (1995-2001)

The United Nations Conference on Environment and Development in 1992, held in Rio de Janeiro, was the initial cornerstone of recognizing POPs on the global level. In Rio, governments from over 140 nations recommended taking measures to reduce and eliminate discharges of persistent organic pollutants for the first time. Delegates at the Summit developed a set of principles to protect the environment and urge sustainable development of the nations, called the Rio Declaration,[1] and also adopted a detailed blueprint for actions to achieve sustainable development in the coming century, called Agenda 21.[2]

Chapter 17 of Agenda 21[3] deals with the protection of the oceans; all kinds of seas, including enclosed and semi-enclosed seas; coastal areas; and the protection, rational use and development of their living resources. Section B of Chapter 17.28[4] identifies the need to reduce and eliminate emissions and discharge of organohalogens and other persistent organic pollutants, stating:

As concerns other sources of pollution, priority actions to be considered by States may include:

(d) Eliminating the emission or discharge of organohalogens compounds that threaten to accumulate to dangerous levels in the marine environment;

(e) Reducing the emission or discharge of other synthetic organic compounds that threaten to accumulate to dangerous levels in the marine environment;

(f) Promoting controls over anthropogenic inputs of nitrogen and phosphorus that enter coastal waters where problems, such as eutrophication threaten the marine environment or its resources;

(g) Cooperating with developing countries, through financial and technological support, to maximize the best practicable control and reduction of substances and wastes that are toxic, persistent or liable to bio-accumulate and to establish environmentally sound land-based waste disposal alternatives to sea dumping;

Promoting the use of environmentally less harmful pesticides and fertilizers and alternative methods for pest control, and considering the prohibition of those found to be environmentally unsound.

---

1   Rio Declaration on Environment and Development, *at* http://www.unep.org/unep/rio.htm (last visited on Jan. 25, 2002).

2   UN Conference on Environment and Development, Agenda 21, *at* http://un.org/esa/sustdev/agenda21text.htm (last visited on Jan. 25, 2002).

3   *Id. at* ch. 17, *at* http://www.unep.org/Documents/Default.asp?DocumentID=52. (last visited on Jan. 24, 2002) (Protection of the Oceans, all kinds of seas, including enclosed and semi-enclosed seas and coastal areas and the protection, rational use and development of their living resources).

4   *See id.* ch. 17(28).

Chapter 19 of Agenda 21 deals with the "Environmentally Sound Management of Toxic Chemicals Including Prevention of Illegal International Traffic in Toxic and Dangerous Products."[5] Chapter 19.21 recommends:

> (a) Strengthening research on safe/safer alternatives to toxic chemicals that pose an unreasonable and otherwise unmanageable risk to the environment or human health and to those that are toxic, persistent and bio-accumulative and that cannot be adequately controlled.

Other relevant sections of Chapter 19 are:

> 19.45. In the agricultural area, integrated pest management, including the use of biological control agents as alternatives to toxic pesticides, is one approach to risk reduction.

> 19.46. Other areas of risk reduction encompass the prevention of chemical accidents, prevention of poisoning by chemicals and the undertaking of toxic vigilance and coordination of clean-up and rehabilitation of areas damaged by toxic chemicals.

Further actions to achieve environmentally sound management of toxic chemicals are also suggested on Chapter 19.49 (a) and (c), which state:[6]

> Activities
>
> (a) Management-related activities
>
> 19.49. Governments, through the cooperation of relevant international organizations and industry, where appropriate, should:
>
> (a) Consider adopting policies based on accepted producer liability principles, where appropriate, as well as precautionary, anticipatory and life-cycle approaches to chemical management, covering manufacturing, trade, transport, use and disposal;
>
> (c) Adopt policies and regulatory and non-regulatory measures to identify, and minimize exposure to, toxic chemicals by replacing them with less toxic substitutes and ultimately phasing out the chemicals that pose unreasonable and otherwise unmanageable risk to human health and the environment and those that are toxic, persistent and bio-accumulative and whose use cannot be adequately controlled.

The motivation for the establishment of the Stockholm Convention was to bring into equilibrium the needs of the developing world to produce and export agricultural products to the developed world on one hand and, on the other hand, to address developed country concerns over pesticide residue in foods imported from the developing world using the very chemicals banned at home.[7]

---

5   *See id.* ch. 19.
6   *See id.* ch. 19.49(a) & (c).
7   WHO, Division of Health and Environment, *Pesticides and Health in the Americas*, 1993 Envtl. Series No. 12, at 15 (Feb. 1993).

A. Preparatory Work for the Intergovernmental Negotiating Committee on POPs

Focusing on these recommendations and the historical foundation of international cooperation on chemicals at the technical level,[8] UNEP Governing Council at its ninth meeting on May 25th 1995,[9] first decided to assess 12 persistent organic pollutants, and then to undertake international action to prepare an appropriate international legal mechanism on POPs. At that UNEP Governing Council meeting, Decision 18/32 was adopted, inviting:

> the Inter-Organization Program for the Sound Management of Chemicals, working with the International Program on Chemical Safety, and the Intergovernmental Forum on Chemical Safety, with the assistance on an appropriate ad hoc working group, to initiate an expeditious assessment process, initially beginning with the short list of persistent organic pollutants that is currently being discussed by the United Nations Economic Commission for Europe in the context of the Convention on Long-rage Transboundary Air Pollution, adopted in Geneva on 13 November 1979.[10]

The twelve persistent organic pollutants referred to herein are: aldrin, dieldrin, endrin, mirex, DDT, chlordane, heptachlor, toxaphene, hexachlorobenzene, PCBs, dioxins and furans. The work first considered these initial twelve listed POPs because of the uncontroversial acceptance of the risk these substances, leaving aside more controversial and potentially more difficult substances to be negotiated later. At the time, the twelve chemicals were designated as the "dirty dozen."[11]

The need to further develop an international legally binding treaty to address POPs required a strong international agency that would act as a coordinator and be able to catalyze governments to act. UNEP was the logical choice as the lead environmental agency of the United Nations system. Additionally, UNEP had expertise on chemicals since the establishment of the International Register of Potentially Toxic Chemicals in 1976.[12]

Before POPs negotiations started, public health issues related to them were soon identified. UNEP recognized the need to coordinate any further action with the World Health Organization.[13] It was also necessary to identify a robust scientific foundation to strengthen the mandate to negotiate the POPs agreement.

---

8   Including the 1985 International Code of Conduct on the Distribution and Use of Pesticides (FAO); 1989 London Guidelines for the Exchange of Information on Chemicals in International Trade (UNEP); 1992 Rio Summit, Agenda 21 at chapter 17 and 19 (UNEP); 1994 Code of Ethics on the International Trade in Chemicals (UNEP) to name a few.

9   *See id., supra*, note 259.

10   *See id.*

11   John Buccini, *Persistent Organic Pollutants: Recent Developments in the Intergovernmental Forum on Chemical Safety*, at http://www.chem.unep.ch/pops/POPs_Inc/proceedings/stpetbrg/buccini.htm (last visited on Jan. 25, 2002).

12   *See Database for Global Environment Research*, at http://www-cger.nis.go.jp/cger-e/db/info-e/InfoDBWeb/db/irptc.htm (last visited on Jan. 24, 2002).

13   The use of DDT in the fight against malaria in developing countries is a WHO major concern.

For that reason, the International Forum on Chemical Safety (IFCS) was convened in Manila in 1996,[14] bringing experts to consider the science on POPs and prepare a recommendation to the UNEP Governing Council on a future global treaty on POPS. The IFCS experts prepared two reports in June and July 1996, presenting to the UNEP Governing Council sufficient scientific data to justify negotiation of a global legally binding international treaty on persistent organic pollutants.[15] Another important recommendation of the report was to concentrate the focus on persistent organic pollutants as opposed to every chemical substance.[16]

As a preparatory measure for the negotiations of a global legally binding treaty, UNEP Governing Council Decision 19/13C of February, 1997,[17] requested in Section 6 (c):

> coordination among different regional and international initiatives on persistent organic pollutants to ensure harmonized environmental and health outcomes from mutually supportive and effective programs that result in the development of policies with complementary, and non-conflicting, objectives and that avoid overlap and duplication with other international and regional conventions and programs;
>
> (c) need for capacity-building in countries and region.

Thereafter, UNEP Chemicals promoted a series of eight workshops to build awareness and technical understanding of all the issues related to POPs. These POPs workshops were held in the following countries:[18]

a) Saint Petersburg, Russia, in July 1997;

b) Bangkok, Thailand, in November 1997;

c) Bamako, Mali, in December 1997;

d) Cartagena, Colombia, in January 1998;

e) Lusaka, Zambia, in March 1998;

f) Iguazu, Argentina, in April 1998;

g) Ljubljana, Slovenia, in May 1998, and

h) Abu Dhabi, United Arab Emirates, June 1998

---

14   Mandate for this IFCS Experts Meeting is on UNEP Governing Council Decision 18/32, *supra* note 259.

15   *See* IFCS Expert's Meeting on POPs, *Problems with Persistent Organic Pollutants Towards a Better Alternatives*, 1996 to the UNEP Governing Council, *at* http://irptc.unep.ch/pops/indxhtms/ifcsall.html (last visited on Jan. 24, 2002).

16   *Intergovernmental Forum on Chemical Safety, at* http://irptc.unep.ch/pops/indxhtms/manwg6.html (last visited on Jan. 24, 2002).

17   UNEP Governing Council Decision 19/13, *at* http://www.unep.org/Documents/Default.asp?DocumentID=96&ArcticleID=1470.

18   Final Report, John Buccini, Chair of the IFCS Working Group on POP, 3rd Meeting of the Intersessional Group Intergovernmental Forum on Chemical Safety, *at* http://www.who.int/ifcs/isg3/d98-21b-en.htm (last visited on Jan. 24, 2002).

An array of information was presented at these workshops relevant to POPs, including scientific, technical, economic and social issues that offered participants grounds to believe that it was time to proceed with a multilateral action on POPs.

One important multilateral treaty which contained a section on relevant POPs was the 1982 United Nations Convention on the Law of the Sea. Article 207, section 5 of part XII of this treaty sets out provisions regarding pollution from land-based sources. It recommends state parties to the convention act, especially through competent international organizations or diplomatic conferences, to employ efforts to establish global and regional rules, standards, and recommended practices and procedures to prevent, reduce and control pollution of the marine environment.[19] All rules, standards and recommended practices and procedures referred to above "shall include those designed to minimize, to the fullest extent possible, the release of toxic, harmful or noxious substances, especially those which are persistent, into the marine environment."[20] This language of the Law of the Sea treaty related to POPs in the marine environmental only, was the first inclusion of such a reference in an international legally binding instrument.

Inspired by this Article 207 of the Law of the Sea treaty and by UNEP Governing Council's Decision 18/31,[21] which requested support for the establishment of a Global Programme of Action for the Protection of the Marine Environment from Land-based Activities,[22] a decision was made to hold a conference in Washington, DC on November 3 1995, to look into the "Protection of the Marine Environment from Land-Based Activities." The conference, sponsored by UNEP and hosted by the United States, sought to address the problem of marine pollution originated by land-based activities. The result of this Conference was the adoption of a Global Program of Action[23] with the consensus of over 109 governments that attended the meeting, including the United States.[24] This Global Program of Action included a recommendation for the preparation of an international legally binding treaty for the reduction and eventual phase-out of the "dirty dozen" chemicals pursuant to paragraph 17 of

---

19   United Nations Convention on the Law of the Sea, *at* http://www.un.org/Depts/los/convention_agreements/texts/unclos/closindx.htm (last visited on Jan. 28, 2002).
20   *See id.* art. 212.
21   UNEP Governing Council Decision 18/31, The Protection of the Marine Environment from Land-Based Activities, *at* http://chem.unep.ch/pic/gc13-31.html (last visited on Jan. 25, 2002).
22   UNEP, Global Programme of Action for the Protection of the Marine Environment from Land-Based Activities, UN Doc. UNEP/OCA/LBA/IG.2/7 (1995), *at* http://www.unep.org/unep/gpa/ga51-189.htm (last visited on Jan. 25, 2002).
23   *See id.*
24   National Ocean Service, *International Program Office Report on the Global Program of Action for the Protection of the Marine Environment from Land-Based Activities*, *at* http://international.nos.noas.gov/conv/gpa.html (last visited on Jan. 25, 2002).

the "Washington Declaration on Protection of the Marine Environment from Land-Based Activities,"[25]

> acting to develop, in accordance with the provisions of the Global Programme of Action, a global, legally binding instrument for the reduction and/or elimination of emissions, discharges, and where appropriate, the elimination of the manufacture and use of the *persistent organic pollutants* identified in decision 18/32 of the Governing Council of the United Nations Environment Programme.[26]

Another important preparatory event took place at the second Inter-Sessional Group (ISG2) meeting of the IFCS, held in Canberra, Australia in March 1996, which approved a report previously assigned by the UNEP Ad Hoc Working Group on POPs. This working group was established to implement Decision 18/32 (a), (c) and was formed by representatives from intergovernmental organizations, governments, industry, public interest groups and scientific organizations from all over the world.[27]

The final report of ISG2 demonstrated that:[28]

> a) the amount of available scientific evidence on the chemistry, toxicology, transport pathways, origin, transport and disposal on a global scale constituted sufficient evidence to support a decision to negotiate a international legally binding document on the initial list of the 12 POPs in order to reduce the risks to human health and the environment;

> b) enough evidence was presented to verify that the 12 POPs include pesticides, industrial chemicals and unintentionally produced by-products and contaminants;

> c) further research should be carried out to address socio-economic issues, production and use of POPs around the world and identification of alternative products and technologies to POPS at a future experts meeting, which should be held in Manila in 1996; and

> d) further assessment needs to be conducted to address task (e) of Decision 18/32 and "assess realistic response strategies, policies and mechanisms for reducing and/or eliminating emissions, discharges and losses of persistent organic pollutants."[29]

---

25   Washington Declaration on the Protection of the Marine Environment from Land-Based Activities, *at* http://www.unep.org/unep/gpa/pol2b12.htm (last visited on Feb. 25, 2002).
26   *See id.* ¶ 17.
27   International Forum on Chemical Safety Experts Meeting, Final Report at UNEP Doc. IFCS/EXP.POPs.2 (June 17-19, 1996), *at* http://irptc.unep.ch/pops/indxhtms/manpops2.html (last visited on Jan. 25, 2002).
28   IFCS Experts Meeting on POPs: Final Report: Background Mandate, ¶ 5, *at* http://irptc.unep.ch/pops/indxhtms/manexprr.html (last visited on Jan. 25, 2002).
29   *See supra* note 259, at 1(e).

On June 17th to 19th, 1996, the IFCS[30] Experts Meeting on Socio-Economic Considerations for a Global Action on POPs[31] was held in the city of Manila, co-sponsored by Canada and the Republic of the Philippines. Once again enough evidence was present to recommend immediate efforts to prepare an international legally binding instrument addressing POPs. IFCS also concluded that the participation of developing countries in responding to international action on POPs is essential since they are users of great amounts of pesticides. The experts agreed that special attention should be given to assist developing countries to meet their needs, including: training workers and trainers; information exchange; strengthening national legislation and enforcement capabilities; disposal capability; research facilities; capacity building; public awareness of alternatives to POPs and alternative technologies; and public awareness of hazards of POPs.[32] The Forum's recommendation included data on human adverse effects of POPs and the urgent need to reduce and eliminate discharges at national, regional and international levels. The need was also pointed out to develop science-based criteria in order to add other persistent pollutants to the initial list of twelve.[33] The Forum also pointed out that most industrial states have robust domestic legislation that can regulate all aspects of POPs, including import, export, manufacture, storage, use, release, disposal, and transport, whereas the great majority of developing nations do not have the legal framework or even the administrative structure required to address toxic chemicals appropriately.[34]

B. First INC Meeting—Montreal, 29 June-3 July 1998

As a result of the recommendations from the conference, working group, and experts group described above, the UNEP Governing Council, meeting in Nairobi, adopted a resolution, Decision 19/13C, requesting UNEP to establish an Intergovernmental Negotiating Committee (INC)[35] to draft an international legally binding treaty to eliminate, initially, the twelve POPs. At this meeting the final report from IFCS was formally adopted by the UNEP Governing Council.[36] It was

---

30 IFCS was established on April 29th, 1994 at the International Conference on Chemical Safety, held in Stockholm, Sweden and their Website is: http://www.who.int/ifcs/ (last visited on Jan. 25, 2002).

31 Experts Meeting on Socio-Economic Considerations for a Global Action on POPs, UN Doc. IFC/EXP.POPs. (May 2-28, 1996), *at* http://irptc.unep.ch/pops/indxhtms/manpops2.html (last visited on Jan. 25, 2002).

32 Final Report of IFCS Working Group on POPs, *at* http://irptc.unep.ch/pops/indxhtms/manwgrp.html (last visited on Jan. 25, 2002).

33 Experts Meeting Final Report, IFCS/EXP.POPs./Report.1.Final (June 20, 1996). http://irptc.unep.ch/pops/ifcsvarious.html (last visited on Jan. 25, 2002).

34 G. Heaton, R. Banks & D. Ditz, Missing Links: Technology and Environmental Improvements in the Industrializing World (The World Resources Institute 1994).

35 UNEP Governing Council Decision 19/13, *International Action to Protect Human Health and The Environment Through Measures Which Will Reduce and/or Eliminate Emissions and Discharges of Persistent Organic Pollutants, Including the Development of an International Legally Binding Instrument*, *at* http://irptc.unep.ch/pops/gcpops_e.html (last visited on Jan. 25, 2002).

36 *See id.* § 1.

conclusive regarding the need to reduce the risks to human health and the environment arising from the release of POPs.

The drafting of an international legally binding treaty was on its way. At every INC participants selected both a chair and a bureau of the plenary, and developed the basic Rules of Procedure to guide the works. Due to the complexity of the theme and the degree of expertise required, the plenary established working groups and negotiating groups on specific issues, including an expert group, known as the "Criteria Expert Group," to develop procedures and criteria to add new chemicals to the final text.[37] The plenary also established a working group entitled the "Implementation Aspects Group" (IAG) to study and propose solutions as for funding, technology transfer, technical assistance, and other relevant items regarding the implementation of the treaty.[38] Another important strategy adopted to expedite negotiations was the establishment of regional working groups, such as the Group of 77 and China, European Union countries, the Group of Latin American and Caribbean countries (GRULCA) and the JUSSCANNZ countries.[39] There was also broad participation by NGOs[40] and intergovernmental organizations.[41] UNEP Chemicals, serving as the secretariat of the INCs, provided all logistical and operational support. Other United Nations specialized agencies contributed greatly, providing detailed information in the field of their expertise. For example, FAO provided information on its program to enable the safe destruction of POPs and other chemicals in developing countries the World Health Organization helped with all the issues of public health, with particular input on the use of DDT for malaria control and GEF shared its experience in funding for international treaties.[42]

Main Issues Negotiated

At the negotiation, the need for a solid technical and scientific basis for work was soon clearly understood by delegates as a means of generating awareness

---

[37] Report, The Intergovernmental Negotiating Committee for an International Legally Binding Instrument for Implementing International Action on Certain Persistent Organic Pollutants, 1st Sess., UN Doc. UNEP/POPS/INC.1/7 (July 3, 1998), *at* http://www.chem.unep.ch/sc/documents/meetings/inc1/inc1finalreport-e.htm (last visited on Jan. 28, 2002).

[38] *See id.*

[39] JUSSCANNZ Countries Group was formed by Japan, the U.S., Switzerland, Canada, Australia, Norway, New Zealand, and Korea.

[40] NGOs such as: WWF, Sierra Club, Pesticide Action Network, International POPs Elimination Network (IPEN), Greenpeace International, Circumpolar Conservation Union (CCU), American Chemistry Council (ACC), European Chemical Industry Council (CEFIC), International Council of Chemical Associations (ICCA), World Chlorine Council (WCC), etc.

[41] Intergovernmental Organization such as European Commission, The World Conservation Union (IUCN), Organization for the Prohibition of Chemical Weapons (OPCW), etc.

[42] UNEP, 1st Sess., UN Doc. UNEP/POPS/INC.1/INF/7 (1998), *at* http://irptc.unep.ch/pops/POPs_Inc/INC_1/inf7.htm (Report by the World Health Organization).

and sounding the alarm to the world.[43] Therefore, there was an agreement among participants to this first INC to establish an expert group to determine criteria and the procedure for identifying additional POPs as candidates for future action.[44] This experts group had to develop such criteria on a sound scientific basis but, also, they agreed on the need to apply the precautionary principle in the procedure for identifying and controlling POPs.[45] The first INC then decided unanimously to set up a group with a mandate to prepare the terms of reference[46] for this expert group to develop science-based criteria and a procedure for identifying additional POPS as candidates for future international action.[47] The group was formed by experts appointed by governments and represented all regions of the world. The participation of NGOs and Inter Governmental Organizations (IGO) as observers also was included.[48] Their work had to concentrate on the establishment of criteria pertaining to persistence, bioaccumulation, toxicity and exposure in different regions, and had to take into account the potential for regional and global transport, including dispersion mechanisms for the atmosphere and the hydrosphere, migratory species, and the need to reflect possible influences of marine transport and tropical climates.[49]

The second concern at the first INC was the establishment of a financial assistance mechanism to help developing countries meet their obligations arising from the international legally binding instrument on POPs.[50] Activities such as POPs monitoring, compiling inventories, capacity-building, identifying and obtaining alternatives to POPs, and destroying or disposing of obsolete stocks in an environmentally sound manner are costly and require a great deal of expertise.[51] Based on the experience of the Montreal Protocol on Substances that Deplete the Ozone Layer, participants believe that its success was largely due to its robust financial mechanism. The INC requested its Secretariat to prepare a report for consideration at its second meeting on all existing programs that provide financial and technical assistance with regard to the management and elimination of chemicals. For the purpose of this report, the experience of organizations such as Global Environment Facility, it was decided that the input of the World Bank, regional development banks and member organizations of IOMC (UNEP, FAO, WHO, ILO, UNIDO, OECD, and UNDP) should be taken into

---

43 UNEP, 1st Sess., UN Doc. UNEP/POPS/INC.1/INF/2 (1998), *at* http://irptc.unep.ch/pops/POPs_Inc/INC_1/inf2.htm (Preparation of an international legally binding instrument for implementing international action on certain persistent organic pollutants).
44 *See id*. at § 27(a).
45 *See id*.
46 *See id*. at Annex II (Terms of Reference for the Criteria Expert Group for POPS).
47 *See id*. § 64(a).
48 *See supra*, note 390 at Annex II, ¶ 4 (Participation).
49 *See id*. ¶ 2.
50 *See* IOMC, *supra* note 266, at § 30(a).
51 *See id*.

consideration.[52] The INC decided to create a subsidiary body to examine implementation aspects of POPs, with a special recommendation to look into two broad topics: a) identify areas of need that would require technical and financial assistance, and b) identify possible existing or future sources of technical and financial assistance.[53] Their work was to begin at the second session of the INC.

The third most hotly debated topic among participants at the first INC was whether the international legally binding instrument should only regulate international trade in POPs, or if all international trade should be banned, except for the purpose of their destruction.[54] A fourth hot issue present at all negotiations was the wording of the "objective" of the Convention. The question was whether it should aim at elimination or merely at management of chemical toxic substances.[55] This session was ended with the recommendation that the Secretariat should develop a draft text of the convention that would build upon comments and suggestions made at this first INC and any comments sent to the Secretariat by September 1, 1998. This text would then provide the basis of discussion for the next INC meeting.

C. First "Criteria Experts Group" Meeting—Bangkok, October 1998

As previously mentioned on pages 169 and 170, before the second INC meeting and pursuant to UNEP Governing Council Decision 19/13C,[56] paragraph 9, adopted on February 7th, 1997, a Criteria Experts Group[57] for persistent organic pollutants was created to look into science-based criteria and procedures to identify and add other POPs to the international legally binding instrument. The text of the decision recognizes:

> ... the need to develop science-based criteria and a procedure for identifying additional persistent organic pollutants as candidates for future international action and requests the intergovernmental negotiating committee to establish, at its first meeting, an *expert group* to carry out this work....[58]

The first meeting of the Criteria Expert Group (CEG) was held in Bangkok[59] from 26th to 30th October, 1998, at the invitation of the Government of Thailand, at the headquarters of the United Nations Economic and Social Commission for Asia and the Pacific (ESCAP). Government-designated experts from all over the

---

52 See id. 62(a).
53 See id. 59(a), 63(a).
54 See id. 32(a).
55 See id. 35(a).
56 See Intergovernmental Negotiating Committee, *supra* note 381.
57 See Intergovernmental Negotiating Committee, *supra* note 381, at Annex II (Criteria Expert Group 1, Terms of Reference for its work).
58 See Intergovernmental Negotiating Committee, *supra*, note 381, ¶ 9.
59 Final Report of the Criteria Experts Group 1, UN Doc. UNEP/POPS/INC/CEG/1/3, *at* http://www.chem.unep.ch/sc/documents/meetings/ceg1/CEG1-3.htm (last visited on Jan. 28, 2002).

world attended this first meeting.[60] Participation of NGOs[61] was also of great importance, especially in sponsoring a number of international meetings on different topics such as national efforts to phase out DDT; effects of endocrine disruption and presentation of films and documentaries, including those on POPs contamination of indigenous populations.[62] Due to the complexity of the problem a conclusive report, however, was not achieved. Recommendations were adopted to take further action on gathering information and scientific data on the following points:[63]

a) Alternatives to POPs, including their cost, efficacy, risks and availability;

b) Pros and cons of these alternatives, possibly delineating the effects on health, agriculture, biota, economy of the nation, and how their alternatives would help countries move towards sustainability;

c) Detailed measures to deal with waste and disposal of POPs, particularly, obsolete stockpiles.

D. Second INC Meeting—Nairobi, 25-29 January 1999

A second meeting of the INC was held at the UNEP headquarters in Nairobi from 25th to 29th, January 1999 with the participation of 104 countries, 5 United Nations specialized agencies, 3 intergovernmental organizations, and over 100 NGOs. By now the theme was of great interest to most countries worldwide and negotiations began to take place.[64]

---

60 Experts from these countries were present: Antigua and Barbuda, Armenia, Australia, Austria, Belarus, Bhutan, Brazil, Cameroon, Canada, Chad, Chile, China, Colombia, Côte d'Ivoire, Croatia, Cuba, Denmark, El Salvador, Fiji, Finland, France, Gambia, Germany, Iceland, India, Indonesia, Iran, Italy, Japan, Kazakhstan, Kenya, Lao People's Democratic Republic, Malawi, Malaysia, Mongolia, Mozambique, Netherlands, New Zealand, Norway, Peru, Philippines, Qatar, Republic of Korea, Russian Federation, Singapore, Slovenia, South Africa, Sweden, Switzerland, Syrian Arabic Republic, Thailand, the former Yugoslav Republic of Macedonia, Ukraine, Britain, USA, Uzbekistan and Zimbabwe.

61 NGOs present at the first Experts Group Meeting: Asian Environmental Society, Canadian Chemical Producer's Association, Commonwealth Science Council, European Center for Ecotoxicology and Toxicology of Chemicals, European Chemical Industry Council, Global Crop Protection Federation, Greenpeace International, Indian Chemical Manufacturers Association, International Confederation of Free Trade Unions, International Council of Chemical Associations, International Council of Environmental Law, Pesticide Action Network–Asia and the Pacific, World Chlorine Council, and the World Wide Fund for Nature.

62 Telephone interview with Claudio Torres Nachón, LLM, Director of the Center of Environmental Law and Economic Integration for the South. Landmine Monitor Mexico, who attended all INCs. E-mail: dassurct@prodigy.net.mx.

63 Report of the first session of the Criteria Expert Group (CEG1) for POPs, Bangkok, (Oct. 26-30, 1998), at http://www.chem.unep.ch/sc/documents/meetings/ceg1/CEG1-3.htm (last visited on Jan. 28, 2002).

64 II INC Meeting, Report of the INC for an International Legally Binding Instrument for Implementing International Action on Certain Persistent Organic Pollutants on the Work of Its Second Session, UN Doc. UNEP/POPS/INC.2/6, at http://www.chem.unep.ch/sc/documents/meetings/inc2/final_report?INC2-6finrep-eng.html (last visited on Jan. 28, 2002).

Chapter 4

At the first session of the Intergovernmental Negotiating Committee, the Secretariat was instructed to prepare a report pertaining to technical and financial assistance[65] for analysis at this second session of the INC.[66] In response to that request, the Secretariat concluded that significant technical and financial assistance was potentially available,[67] since the Global Environmental Facility was considering financing projects consistent with the "Outline of a UNEP project portfolio on persistent organic pollutants" and had preliminarily allocated US$15 million for it.[68] Another available source was the United Nations Development Programme that sponsors activities to promote reduction of pesticide use and waste minimization through innovative means.[69] Finally the World Bank also supports the development of environmental institutions, including those involved in pollution control, as part of its environmental portfolio of nearly US$ 11 billion.[70]

Many government representatives from developed nations in particular at the Second INC reported on activities in their countries dealing with POPs. The Committee recommended the preparation of a master list of actions on the reduction and/or elimination of POPs in order to avoid duplication of efforts, to ensure efficient use of resources, and to facilitate coordination and cooperation among nations at national, regional and international levels. UNEP Chemicals reported to the INC that some information was already available through letters and questionnaires sent to governments and organizations and these responses should be used to prepare an up-to-date information list. Under UNEP Chemicals authority, these actions were reported:

a) regional and sub-regional awareness raising and management workshops;[71]

b) Guidelines for the Identification of PCBs and Materials Containing PCBs and Inventory of world-wide PCB destruction capacity;[72]

---

65 Report by the Secretariat at the II INC, *Existing mechanisms for providing technical and financial assistance to developing countries and countries with economies in transition for environmental projects*, UNEP/POPS/INC.2/INF/4, *at* http://www.chem.unep.ch/sc/documents/meetings/inc2/en/inc2-INF4.htm (last visited on Jan. 28, 2002).

66 See id.

67 UNEP, Report by the Secretariat, *Existing Global, Regional and Bilateral Programmes Providing Technical and Financial Assistance with Regard to the Management and Elimination of Chemicals*, UN Doc. UNEP/POPS/INC.2/5 (1999), *at* http://www.chem.unep.ch/sc/documents/meetings/inc2/en/inc2-5.htm (last visited on Jan. 28, 2002).

68 UNEP, Outline of a UNEP project portfolio on persistent organic substances, UN Doc. UNEP/POPS/INC.1/INF/14 (1999), *at* http://www.chem.unep.ch/sc/documents/meetings/inc1.htm.

69 See id. at 12.

70 See id. at 13.

71 Workshops Web Sites: UNEP-IFCS Awareness raising and UNEP POPs Management, *at* http://www.chem.unep.ch/sc/documents/background/proceedings/default.htm (last visited on Jan. 29, 2002).

72 UNEP, *Guidelines for the Identification of PCBs and Materials Containing PCBs, at* http://chem.unep.ch/pops/newlayout/repdocs.html (last visited on Jan. 29, 2002).

c) information on POPs alternatives and approaches to replace and/or reduce the releases of POPs chemicals;[73]

d) Global Survey on POPs;[74]

e) Inventory of Information Sources of Chemicals;[75]

At this stage of negotiations the issue of whether POPs should be eliminated or regulated was especially difficult to negotiate because of the differences that emerged among representatives of governments, environmental and industry-affiliated organizations. The Committee decided to leave this issue to the next meeting when interested parties would suggest better wording.[76] A second major difference involved the debate among participants relating to the addition of other POPs to the text of the Convention. So far broad consensus had made possible the limitation of the text to the initial 12 substances, since they were heavily regulated and/or banned in many countries and no longer protected by patents. Participants requested the Criteria Experts Group to prepare a second meeting to look carefully into criteria for adding new substances to the international legally binding instrument. The Committee agreed and decided to wait for a report at the third INC.[77] In establishing technical and financial assistance to developing countries, the committee agreed to further consider the issue in three different areas: a) identify areas of need that could require technical assistance; b) examine potential costs associated with such assistance, and c) identify possible existing or future sources of technical and financial assistance.[78]

Participants agreed that conducting POPs inventories would be an essential step in implementing an international instrument on POPs and, despite the willingness of developing countries to prepare their POPs inventory, they could not do so because they needed technical and financial assistance immediately.[79] Another important action which was discussed was to establish focal points in each country to help with the organizational structure for the implementation of the Stockholm Convention, to help government bodies achieve their commitments requested by the Conference of the Parties in the future, to channel technical and other kinds of assistance to the rest of the country, and to serve as a feedback mechanism to the Secretariat on the status of implementation of the instrument.[80] Both proposals were accepted and were included in the final text of the Convention.

---

73 UNEP, *Report on information on POPs alternatives and approaches to replace and/or reduce the releases of POPs chemicals,* at http://chem.unep.ch/pops/newlayout/infpopsalt.htm (last visited on Jan. 29, 2002).

74 See id.

75 UNEP, *Inventory of Information Sources of Chemicals,* at http://www.chem.unep.ch/pops/newlayout/infpopschem.htm (last visited on Jan. 29, 2002).

76 See supra note 381, at §§ 49-61.

77 See id. at § 67.

78 See id. at § 78.

79 See id. at § 80.

80 See id. at § 81.

Chapter 4

The issue of free trade in relation to POPs was discussed during the first INC, without the introduction into the draft text of any language at that time. By the second INC, however, an Article N*bis* was proposed by the Government of Australia and the chemical industry through its representatives at the INC, which stated:

> The provision of this Convention shall not affect the rights and obligations of any party deriving from any existing international agreement.

This language was intended to provide for the supremacy of the GATT's dedication to unrestrained trade over any provisions of the draft convention, which overall was intended to prohibit trade in designated chemicals or reduce the use and production of others. Known as the "WTO supremacy clause," this sentence meant that trade rules would govern over environmental or public health concerns.[81] The effort to include this sentence in the treaty failed once environmental groups led by Greenpeace, WWF, and IPEN mobilized opposition. The WTO supremacy clause was not introduced in any subsequent INC, and the final text of the convention makes no reference to it, except in the preamble as a general phrase: "This convention and other international agreements in the field of trade and the environment are mutually supportive."

By the end of the meeting, a preliminary draft text of the Convention was prepared, that contained non-controversial issues such as the role of the Conference of the Parties, Secretariat, amendments to the Convention, right to vote, signature, ratification, acceptance, approval and accession, entry into force and withdrawal.[82] All other clauses were opened to further discussion and proposals, which would be brought before the upcoming INC.

E. Second Criteria Experts Group—Vienna, 14-18 June 1999

UNEP Chemicals, in negotiating the Stockholm Convention, conducted a wide range of meetings and discussions outside the formal meetings of the Intergovernmental Negotiating Committee. These events were important in order

---

81   Article XX of the GATT does provide exceptions to the absolute goal of removing all barriers to free trade. Members of the WTO may enact domestic legislation or adopt regulatory policies to protect public health or the environment, even when this means restrictions on imports or exports of products to the benefit of domestic producers. For example, Art. XX (b) refers to measures to protect human health, animal or plant life or health; and Art. XX (g) refers to the conservation of exhaustible natural resources, if such measures are made effective in conjunction with restrictions on domestic production or consumption. Recent decisions of the WTO dispute resolution panels have recognized the validity of Art. XX exceptions in principle, while declining to apply them to existing cases. *See, e.g.*, Beef Hormones Case (EU Measures concerning meat and meat products (hormones), Jan. 16, 1998, at http://www.wto.org/english/tratop_e/dispu_e/distabase_e.htm); Shrimp-Turtle Case. Some additional transparency may be added to the WTO proceedings by the participation of NGOs in the newly formed Committee on Environment. Until now, most environmental activists have protested the single-minded approach of the WTO to trade as the greatest good regardless of other values and needs. Whether this opposition changes in the future or not remains to be seen.

82   *See supra* note 381, Annex I.

to expedite the process of negotiating and collecting data to serve as foundations for the INC to draft the text of the Convention. In this spirit, a second meeting of the Criteria Expert Group was held in Vienna, 14-18 June 1999,[83] to further discuss the establishment of criteria and procedures to add new chemical substances for future international action. Experts attended the meeting from 63 countries, four United Nations bodies and specialized agencies, four intergovernmental organizations and nine non-governmental organizations.[84]

Their main task was to prepare a draft of science-based criteria and a procedure for identifying additional persistent organic pollutants as candidates for future international action. The Group also presented the following recommendations to be considered by the third INC:

a) Article "F"—listing of substances in Annexes A, B or C.[85]

It was agreed to specify in Annex "A" to the future Convention all chemical substances aimed at elimination of production and use, in Annex "B" all chemical substances for restriction of use and production, and in Annex "C" all unintended industrial by-products. The procedure for choosing which chemical substance goes in which Annex was also resolved by allowing any Party to the Convention to submit a proposal to the Secretariat for listing a given substance. Each party applies the criteria required in Annex D. The Secretariat then verifies whether the admissibility requirements (Annex D) were fulfilled and send the proposal to the Persistent Organic Pollutants Review Committee. The Committee then reviews the proposal and prepares a risk profile in accordance with the rules set forth in Annex "E." Pursuant to the definition agreed on by the Criteria Experts Group, risk profile is

> a comprehensive written review, including analysis and integrated conclusions focused on the scientific information necessary for evaluating whether the substance, as a result of its long-range environmental transport, is likely to lead to significant adverse human health and/or environmental effects.[86]

The Committee then prepares a risk management evaluation, which includes possible control measures for the substance being analyzed, taking into account all social-economic considerations in accordance with Annex "F." A final report recommends inclusion of the substance for listing under the Convention to the

---

83 Terms of Reference of the Criteria Expert Group at their 2nd Sess., Vienna, (June 14-18, 1999), UNEP Doc: UNEP/POPS/INC/CEG/2, *at* http://irptc.unep.ch/pops/CEG-2/meetingdocs/en/ceg2infle.html (last visited on Jan. 29, 2002).

84 Report of the Second Session of the Criteria Expert Group for POPs, UNEP Doc. UNEP/POPS/INC/CEG/2/3, *at* http://irptc.unep.ch/pops/POPs_Inc/INC_3/ceg/en/ceg2rep.htm (last visited on Jan. 29, 2002) (List of attendance of Participants).

85 *See id.* Annex I(A) (art. F: Listing of Substances in Annex A, B, or C).

86 *See id. at* art. F (Definitions).

Chapter 4

Conference of the Parties, who will in turn decide whether to approve the measures proposed by the Committee.

Flow chart for draft proposal on procedures for identifying additional persistent organic pollutants as candidates for future international action

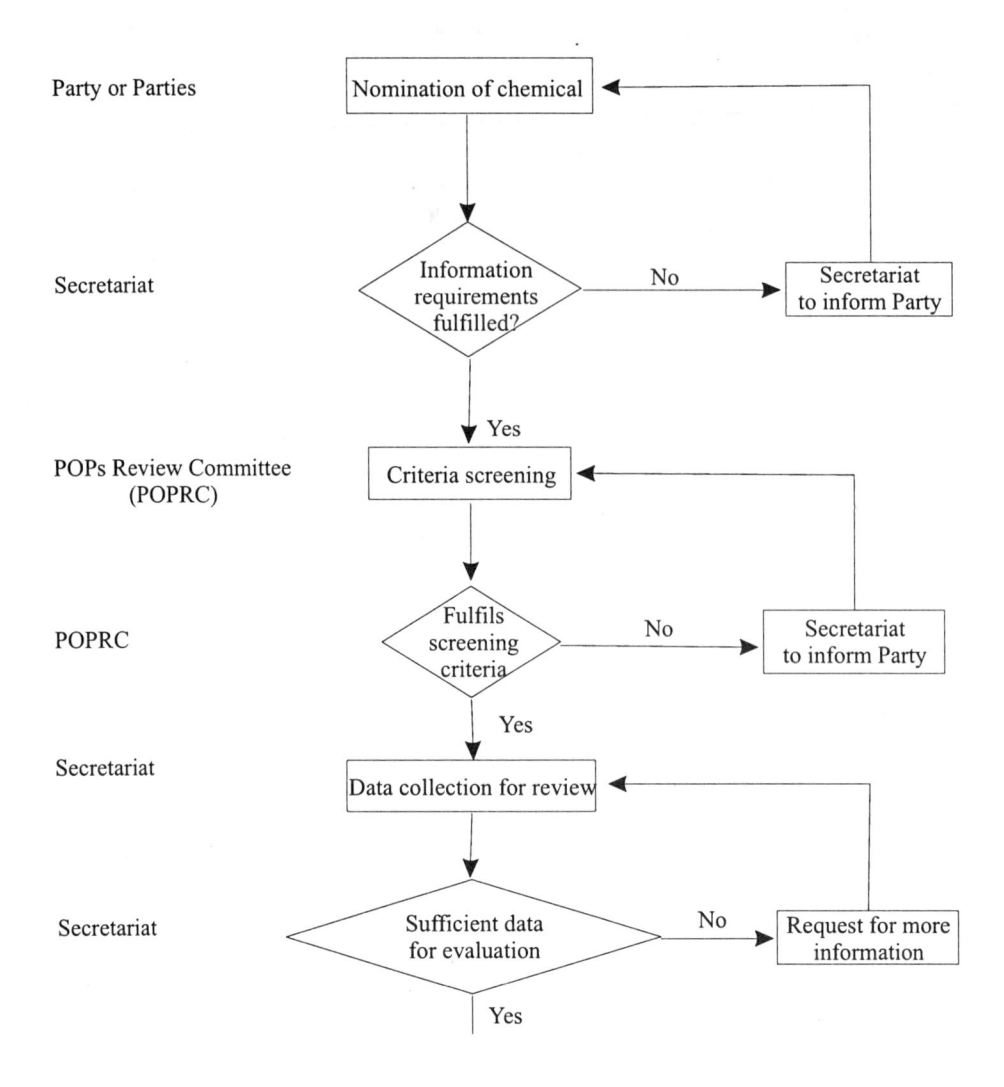

Actor	Step
Party or Parties	Nomination of chemical
Secretariat	Information requirements fulfilled? — No → Secretariat to inform Party
	↓ Yes
POPs Review Committee (POPRC)	Criteria screening
POPRC	Fulfils screening criteria — No → Secretariat to inform Party
	↓ Yes
Secretariat	Data collection for review
Secretariat	Sufficient data for evaluation — No → Request for more information
	↓ Yes

```
                              Continued
                                 │
                                 ▼
POPRC        ┌─────────────────────────────────────┐
             │ Review, preparation of risk profile │
             │ and risk management evaluation,     │
             │ including socio-economic            │
             │ considerations                      │
             └─────────────────────────────────────┘
                                 │
                                 ▼
POPRC              ╱ Recommendation ╲     No      ┌──────────┐
                  ⟨   for listing in  ⟩ ────────→ │ Set aside│
                   ╲  Annex A, B or C╱            └──────────┘
                                 │
                                 ▼
POPRC              ┌──────────────────────┐
                   │ Preparation of Report│
                   └──────────────────────┘
                                 │
                                 ▼
Conference of      ╱   Decision   ╲        No     ┌──────────┐
the Parties       ⟨   to include   ⟩ ───────────→ │ Set aside│
                   ╲ in Annex A,   ╱              └──────────┘
                    ╲  B or C     ╱
                                 │ Yes
                                 ▼
Secretariat        ┌──────────────────────────┐
                   │ Information to all Parties│
                   └──────────────────────────┘
```

———

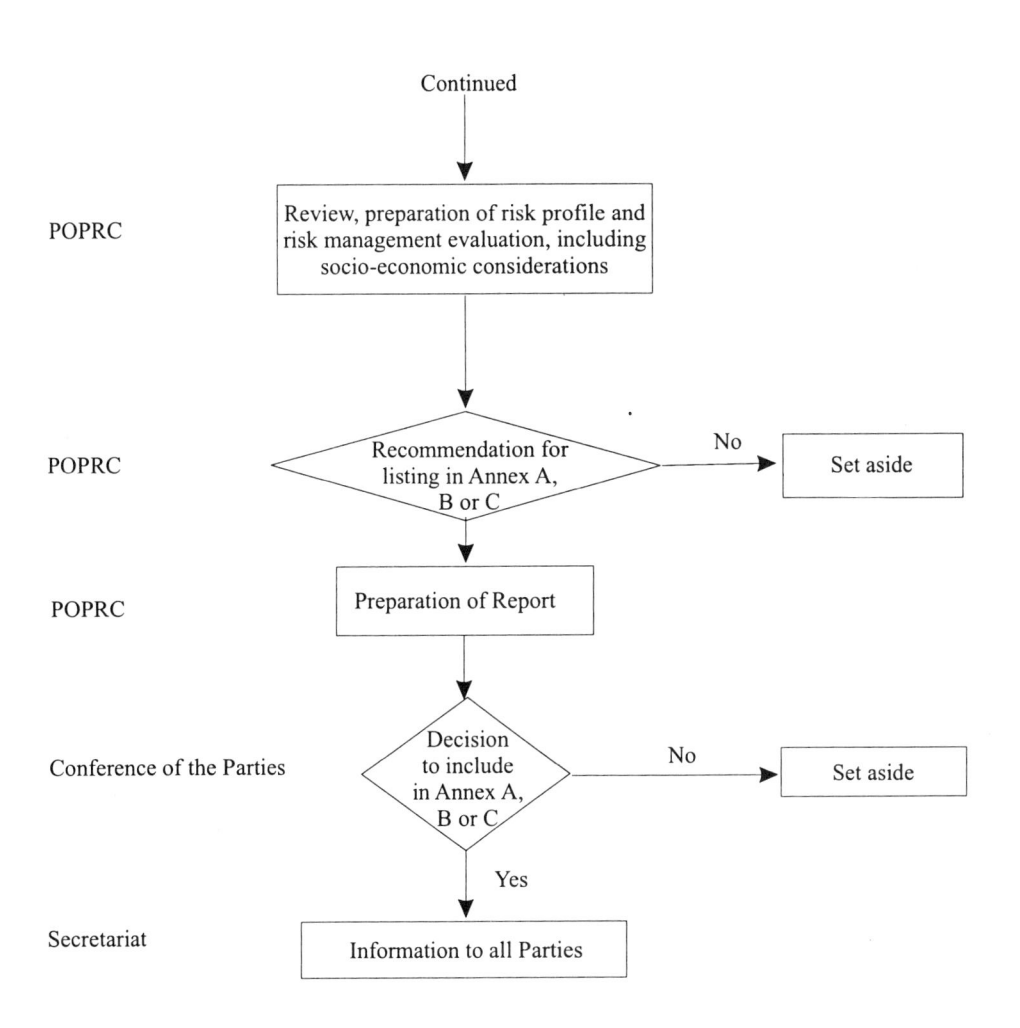

Source: UNEP/POPS/INC/CEG/2/INF/2
http://www.chem.unep.ch/sc/documents/meetings/ceg2/en/ceg2inf2e.html

b) Article "O"—Conference of the Parties[87]

The Conference of the Parties will have the authority to select members of the Persistent Organic Pollutants Review Committee, which will be formed by a limited number of government-designated experts in chemical assessment or management on the basis of equitable geographical distribution among parties from developed and developing countries.

c) Annex "D"—Information Requirements and Criteria for the Proposal and Screening of Proposed Persistent Organic Pollutants[88]

When proposing the listing of a chemical substance in one or more annexes, the following criteria should apply: a) its full identification; b) data on persistence; c) data on bio-accumulation; d) evidence for potential long-range environmental transport, and e) evidence that its toxicity poses the potential for damage to human health or to the environment.

d) Annex "E"—Information Requirement for the Risk Profile[89]

To establish a risk profile of a given substance, information such as its toxicology, sources, chemical and physical properties, persistence, how they are linked to its environmental transport, monitoring data, and status of the substance under international conventions should guide the document in its development.

e) Annex "F"—Information on Socio-Economic Considerations[90]

Under this category, control measures, including management and elimination of a chemical substance, should be proposed, taking into consideration different regions of the world as well as the efficacy and efficiency of the measures, technical feasibility, cost, alternative substances, risk, availability, accessibility, health effects (public, environmental and occupational), agricultural effects, effects on biodiversity, economic aspects, social costs, waste disposal implications, information access and public education, and movements towards sustainable development.

F. Third INC Meeting—Geneva, 6-11 September 1999

The third session of the Intergovernmental Negotiating Committee was held in Geneva, September 6-11th 1999,[91] attended by 115 countries, eleven United Nations bodies and specialized agencies, five intergovernmental organizations,

---

87 See id. at B (Article O: Conference of the Parties).
88 See id. at C (Annex D: Information Requirements and Criteria for the Proposal and Screening of Proposed POPs).
89 See id. D (Annex E: Information Requirements for the Risk Profile).
90 See id. E (Annex F).
91 Final Report of the 3rd INC Meeting—Geneva, (Sept. 6-11, 1999), UNEP Doc. UNEP/POPS/INC.3/4, at http://www.chem.unep.ch/sc/documents/meetings/inc3/inc-english/inc3-4.pdf (last visited on Jan. 29, 2002).

and approximately 100 NGOs.[92] A report[93] was presented to the Committee on existing national legislation and regulatory actions relating to persistent organic pollutants that was prepared by the Secretariat in response to the request made at the 2nd INC Meeting. This information was intended to help the negotiating committee assess the present situation and existing legal structures addressing chemical substances in order to plan future actions to assist countries that have little or no legal mechanisms or update existing legal frameworks on POPs. This first report on countries' existing regulatory mechanisms on POPs was a first attempt to analyze available information. It needs continuous updates. Governments were asked to send information on their existing legislation and regulatory actions in the following categories: chemical substances with a general ban on use, limited bans, bans by the European Community, severe restrictions, and absence of legislation on chemical substances and pesticides. Among the 115 countries that submitted data in response to UNEP Chemicals' request, seventy of them have taken action to ban or severely restrict the use of each of the nine POPs pesticides. Only one country (Sweden) has banned all uses of PCBs and 34 countries have severely restricted its use, being the majority of the industrialized nations. In the great majority of nations responding, there were no actions or very limited actions to control unintentionally generated by-products, such as dioxins, furans and PCBs. Unfortunately, seventy-seven countries did not submit any data, almost all of which were developing countries, including Brazil.[94]

Looking into other international treaties, a report was also presented on lessons learned under the Montreal Protocol[95] in dealing with the issue of phasing out targeted substances. In the same manner that the Montreal Protocol addresses emissions of ozone depleting substances (ODS), lessons can be learned from the actions conducted under the Protocol with respect to emissions of persistent organic pollutants. Collecting data is crucial when preparing later national plans for elimination or reduction of substances. Therefore, clear methodology should be required to collect all the information needed and to make it as accurate as possible. Data should include annual production in most recent years by producer, annual imports, exports, existing stocks and annual use with explanation of the reason for the use.

On monitoring production, imports, and exports on a chemical-by-chemical basis, it is important to note that most countries do not have a harmonized system

---

92 *See id.* B (List of Attendance 9-12).
93 UNEP, Report by the Secretariat, *Compendium of Summary Information on Existing National Legislation Relating to Persistent Organic Pollutants*, UNEP/POPS/INC.3/INF/10, at http://www.chem.unep.ch/sc/documents/meetings/inc3/inf-english/inf3-10/inf10.pdf (last visited on Jan. 29, 2002).
94 UNEP Doc. 3rd INC Sess., UNEP/POPS/INC.3/INF/2, at http://www.chem.unep.ch/sc/documents/meetings/inc3/inf-english/inf3-2/inf3-2.htm (last visited Jan. 29, 2002).
95 UNEP, Report by the Secretariat on Inter-Sessional Work Requested by the Committee, Persistent Organic Pollutants, *Country Strategy Development: Experience and Lessons Learned under the Montreal Protocol*, UNEP/POPS/INC.3/INF/6, at http://www.chem.unep.ch/sc/documents/meetings/inc3/inf-english/inf3-6/inf3-6.pdf (last visited on Jan. 29, 2002).

Chapter 4

for controlling imports and exports. A chemical-by-chemical licensing system was developed under the Montreal Protocol to keep track of import and export of ODS and production data is more easily found since there are only a handful of producers in a particular country. Monitoring POPs can be more difficult because many are sold under different trade names, and special training will be required of customs agents to identify what substance is coming or leaving the country.[96]

Another international treaty analyzed was the Rotterdam Convention on the Prior Informed Consent Procedure for Certain Hazardous Chemicals and Pesticides in International Trade.[97] This instrument is used solely for obtaining and disseminating decisions of importing parties as to whether they wish to receive future shipments of a given chemical substance. It aims to promote shared responsibility and cooperative efforts among parties to the Convention in the international trade of hazardous chemicals for the protection of human health and the environment.[98] Annex III to the Convention lists chemical substances subjected to PIC procedures, among which are the following POPs: aldrin, chlordane, DDT, dieldrin, heptachlor, hexachlorobenzene, and PCBs.[99] Under the Rotterdam Convention, chemical substances in quantities that pose no risk to human health or the environment and are imported for research or analysis are exempt from the information procedure.[100] The Convention does not explain what "quantities that pose no risk to human health or the environment" means, leaving a gap for fraudulent imports.

The placement of the initial list of intentionally produced POPs into the annexes was largely negotiated by starting with the easier ones, placing eight chemicals in Annex A (aldrin, chlordane, dieldrin, endrin, heptachlor, mirex, toxaphene, and hexachlorobenzene), leaving DDT and PCBs out. The United States wanted to introduce a text regarding general exemptions, arguing that exemptions were intended to provide ways to help ensure that the future treaty would be cost-effective and legally viable for as many countries as possible. This proposal reflected U.S. domestic legislation on use of chemical substances for laboratory research, articles in use such as flame retardant clothing containing mirex and products which were in the possession of an end-use consumer like pesticides. The U.S. wanted to have the same standards in the future treaty. Most delegates did not see general exemption as a good approach, but rather preferred to adopt chemical-specific exemptions for public health emergencies.

---

96   *See id.*
97   *See* UNEP, *supra* note 331.
98   *See id.* at art.1 (Rotterdam Convention-Objective).
99   *See id*, at Annex III (Rotterdam Convention).
100 *See id.* art. 3(2)(h) (Scope of the Rotterdam Convention).

G. Fourth INC Meeting—Bonn, 20-25 March 2000

The fourth session of the Intergovernmental Negotiating Committee was held in Bonn, Germany from 20 to 25 March 2000, under the auspices of the German government. Representatives of 122 countries, nine United Nations bodies and specialized agencies, six intergovernmental organizations and over 100 NGOs attended the session. At this fourth session, the plenary of the INC extensively discussed possible wording for Article D—"Measures to reduce or eliminated releases Prohibition of the production and use of certain persistent organic pollutants."[101] Proposals were raised on whether it should emphasize measures to reduce, eliminate, prohibit or restrict the use and production of certain persistent organic pollutants. Scandinavian countries and NGOs were pushing for stricter rules demanding prohibition on their use, production, import and export on one hand, but on the other hand the chemical sector and other industrialized nations were looking into more favorable language allowing continued production and use.

The second controversial issue was adding new substances to the existing listing, with conflicting interests among governments, NGOs and chemical industry representatives. The first draft of Article D stated:

> with the aim of protecting human health and the environment, each party having a regulatory and assessment scheme for new pesticides and industrial chemicals, shall take measures within those schemes to (avoid) (prohibit) (prevent) (regulate) the production (import) (export) and use of the newly developed pesticides and industrial chemicals. . . .

NGOs wanted to prevail with the term "prohibit," but in the end the term "prevent" was adopted and excluded the words "import and export."[102]

As far as the listing of chemicals in Annex 1 (substances aimed at elimination), Annex 2 (restricted use and production), and Annex 3 (unintentional by-products), agreement was reached. Delegates proposed further clear definitions of terms such as "best available techniques," "by-products," and the distinction between "new" and "existing" sources of POPs. Representatives of Latin American NGOs and countries in Asia, Africa and economies in transition asked the fourth INC to establish clear and concrete financial and technical measures to help their countries prepare for dealing with all related issues such as POPs stockpiles, countries' POPs assessment, alternatives to POPs and capacity-building. Canada announced that it would contribute 20 million Canadian Dollars over the next five years, specifically to be used on capacity-building projects in developing countries. The United States then announced that its government would commit US$ 500,000 for regional assessment of persistent toxic substances in the fol-

---

101 UNEP, Report of the Intergovernmental Negotiating Committee for an International Legally Binding Instrument for Implementing International Action on Certain Persistent Organic Pollutants on the Work of its Fourth Session, UNEP/POPS/INC.4/5, *at* http://www.chem.unep.ch/sc/documents/meetings/inc4/eng/inc4-5/en/inc4rpt.doc (last visited Jan. 29, 2002).

102 *See id.*

Chapter 4

lowing year (2000). Japan donated US$ 150,000 to support the work of negotiations of the Stockholm Convention.

At the end of the fourth INC, it was agreed to organize an inter-sessional meeting to discuss and develop financial resources and mechanisms. This small group, comprised of one representative from each of the following countries: Cameron, Canada, China, Colombia, Czech Republic, Denmark, Dominican Republic, France, India, Iran, Japan, Micronesia, Nigeria, Norway, Poland, South Africa, UK, USA, and Uruguay, met in Vevey, Switzerland,[103] June 19-21, 2000, to draft a proposal for an adequate and timely financial assistance mechanism to meet the needs of developing countries so they would be able to implement the future Convention. This work would enable the upcoming INC to negotiate a section on the capacity network concept and financial resources and mechanisms. This inter-sessional group agreed that two Articles should be added to the future treaty, one detailing technical assistance mechanisms and the other on financial provisions. The group recommended that with regard to financial assistance: a) include sources of funds on national, bilateral, regional and multilateral levels; b) establish a body responsible for analyzing each request for funds and follow each case until a decision is make on funding; c) address access to both new and existing funding sources; and d) help out in the development and preparation of funding proposals. This financial mechanism was to include the following attributes, detailed by the group that they be: a) accountable; b) efficient; c) clear; d) transparent; e) accessible, and f) stable.[104]

H. Fifth INC Meeting—Johannesburg, 4-9 December 2001

The 5th Session of the Intergovernmental Negotiating Committee was held in Johannesburg, South Africa, December 4-9, 2001, under the auspices of the South Africa government. Representatives of 121 countries, nine United Nations bodies and specialized agencies, two intergovernmental organizations and over 100 NGOs attended this fifth session of the INC. This meeting was significant in that consensus was reached on all the outstanding issues, allowing the draft convention to be considered and adopted at the subsequent Diplomatic Conference as the final text. In successfully and rapidly designing an international legally binding instrument on POPs, credit should also be given to the process which emerged from a jointly-held value base among governments, NGOs and the chemical industry sector, as more fully explained below in Section E.

Developing countries achieved recognition and response to their needs by working articulately as a group in this negotiating process. A number of representatives of developing countries were instrumental in developing the treaty, namely Ghana, Colombia, Iran, Mexico, Australia, South Africa, to name some but a few. The establishment of a "register of specific exemptions" reflected in

---

103 UNEP, *Report of the Inter-Sessional Meeting on Financial Resources and Mechanisms*, UNEP Doc. UNEP/POPS/INC.5/4 (on file with candidate).

104 *See id*. § 2.5(a)-(m).

Article 4 of the Stockholm Convention[105] is an example of a structure that recognizes differentiated needs of individual countries with respect to the core requirement of setting dates for the elimination of certain chemicals. Each country has its own needs and limitations, and this provision will enable developing nations to identify specific exemptions for each individual substance. This approach also helps in determining what kind of technical assistance a particular country will require and with the identification of alternative products.

In identifying the objectives of the Stockholm Convention, various states proposed alternative language. No agreement had been possible in the earlier INCs, even though there were hundreds of suggestions for language sent to the Secretariat. By the Fifth INC, there were five significant proposals brought to the floor for consideration on the objectives of the Convention:[106]

> 1. Australia: "Promote and support international action to protect human health and the environment through the adoption of measures which will reduce or eliminate emissions and discharges of certain POPs of global concern."

> 2. Norway: "Protect human health and the environment through the reduction and ultimate elimination of releases of POPs in accordance with the precautionary approach."

> 3. USA: "to protect human health and the environment through the reduction and, where feasible, the elimination of releases of POPs of global concern, taking into account technical and socio-economic considerations."

> 4. Portugal: "through applying the precautionary principle, to eliminate POPS so as to protect human health and the environment."

> 5. Venezuela: "to monitor, reduce and subsequently eliminate POPs, including their production, use, import and export, with a view to protecting the environment and human health with the aim of achieving sustainable development."

The final text of Article 1, "Objective," approved at the fifth INC, reads:

> Mindful of the precautionary approach as set forth in Principle 15 of the Rio Declaration on Environment And Development, the objective of this Convention is to protect the human health and the environment from POPS.

The specific language proposed by these five states, therefore, was rejected in favor of a more general approach, that of protecting human health and the environment alone. The stronger reference to elimination of POPs, or even to their reduction, was dropped from this paragraph in its final form due to political opposition and concern about the costs to domestic agriculture and industry.

---

105 *See* The Stockholm Convention, *supra* note 1, art. 4.
106 Report by the Secretariat on the draft of Article B, UNEP/POPS/INC.5/2, *at* http://www.chem.unep.ch/sc/documents/meetings/inc5/En/inc5-2/en/inc5_e.doc (last visited Jan. 29, 2002) (Objectives).

The EU and the G77 plus China were able to include the reference to the precautionary approach (although their preferred language is that of the precautionary principle) over strenuous objection by the United States. The United States managed to change the word "principle" to "approach" as a means of giving the concept a lesser degree of importance or applicability. The text manages to make references to "precaution" and "taking of action in the face of scientific uncertainty" in four places: the preamble,[107] objective,[108] Article 8 on the criteria and process for adding new chemical substances,[109] and Annex C on the list of unintentional produced by-products.[110]

Despite the concern of the major proponents such as Sweden and Canada for language dedicating their efforts to the reduction and elimination of POPs, this article on the objective of the new treaty makes no reference to either. Instead, the concept of reduction and elimination was moved to a different article, Article 3 of the Convention, entitled "Measures to Reduce or Eliminate Releases from Intentional Production and Use" and to Article 5 of the Convention, entitled "Measures to Reduce or Eliminate Releases from Unintentional Production and Use." One additional article, Article 6, refers to reduction and elimination of releases from stockpiles and wastes.

As far as adding new chemicals in the future to the existing list, reliance on a science-based criteria establishes clear procedures with which each country will be able to comply. In the event of a Party to the Stockholm Convention proposing a certain substance for inclusion as one of the POPS, the criteria are purely scientific, using widely accepted methods and biochemical factors.[111] The aim of the Stockholm Convention was to establish science as the basis of compliance for all states parties to the treaty.

Most countries, led by Sweden, agreed on the inclusion of dioxins in Annex A as one of the substances identified for immediate elimination of production and use, but they did not succeed.[112] Dioxins are unintended byproducts of activities such as incineration of wastes, so that countries allowing such activities, even though a minority, successfully protested the inclusion of dioxin in Annex A. An-

---

107 *See* The Stockholm Convention, *supra*, note 1, preamble ("Acknowledging that **precaution** underlines the concerns of all the Parties and is embedded within this Convention.")
108 *See id.* at objectives. ("Mindful of the **precautionary approach** as set forth in Principle 15 of the Rio Declaration on Environment and Development, the objective of this Convention is to protect human health and the environment from persistent organic pollutants.")
109 *See id.* art. 8(7)(a) (". . . **Lack of full scientific certainty** shall not prevent the proposal from proceeding.")
110 *See id.* Annex C(B) ("In determining best available techniques, special consideration should be given, generally or in specific cases, to the following factors, bearing in mind the likely costs and benefits of a measure and consideration of **precaution** and prevention.")
111 *See id.* Annex D (Information Requirements and Screening Criteria).
112 Report of the Intergovernmental Negotiating Committee for an International Legally Binding Instrument for Implementing International Action on Certain Persistent Organic Pollutants on the Work of its Fifth Session, UNEP/POPS/INC.5/7, *at* http://www.chem.unep.ch/sc/documents/meetings/inc5/Eng/finalreport/en/inc5efinrep.doc (last visited Jan. 29, 2002).

nex C was created for substances resulting from "unintentional production", specifically to address the dioxin problem.

Funding also again was heavily negotiated and a decision was only possible at the last minute of the negotiations. However, the text is a victory for developing countries because it provides incremental costs to implement the Convention: "Developed countries shall provide "new" and additional financial resources to enable developing countries to meet their obligations."[113] The Council Meeting of the Global Environmental Facility[114] in November 2000 decided to join in facilitating the implementation of the Stockholm Convention by becoming its financial mechanism. A proposal by the European Union presented at the fourth INC was approved, inviting the Global Environment Facility to become the financial agent to support implementation of the Stockholm Convention.[115] The intersessional group that met in Vevey at the meeting on financial resources and mechanisms recommended to the INC that GEF fulfill the role of the financial mechanism to the Stockholm Convention.[116] They highlighted GEF's previous experience as the financial mechanism for the Convention on Biological Diversity and the United Nations Framework Convention on Climate Change.

The GEF Council approved a draft proposal program, *Draft Elements of an Operational Program for Reducing and Eliminating Releases of Persistent Organic Pollutants in to the Environment*,[117] which is a preparatory operational program to assist with financed activities for an international legally binding instrument for implementing international action on certain persistent organic pollutants, subject to approval by the INC. GEF's catalytic role through full involvement of the three implementing agencies of the POPs treaty, namely the United Nations Development Programme, the United Nations Environment Programme, and the World Bank, would serve to optimize institutional experiences and facilitate all actions to implement the Stockholm Convention. It would also bring new partnerships under the same umbrella already serving as financial mechanisms, including the African Development Bank, Asian Development Bank, European Bank for Reconstruction and Development, Inter-American Development Bank, and other specialized United Nations Agencies such as FAO and UNIDO.[118] Interim financing arrangements were specified in Article 14 of the Stockholm Convention, including the role of the GEF and the voluntary compliance by states prior to the entry into force of the Convention.

---

113 *See* The Stockholm Convention, *supra* note 1, art. 13.
114 Global Environment Facility (GEF) Council Meeting, GEF/ UN Doc. GEF/C.16/6 (Nov. 15, 2000), *at* http://www.chem.unep.ch/sc/documents/meetings/inc5/En/inc5-4/en/inc5_4e.doc.
115 *See* The Stockholm Convention, *supra*, note 1.
116 *See* The Stockholm Convention, *supra* note 1, at 2.6.
117 Council Meeting of GEF, Decision GEF/C.16/6 (2000) (on file with candidate).
118 GEF Report, *Related Work on Persistent Organic Pollutants under the Global Environment Facility*, 5th Sess., UNEP Doc. UNEP/POPS/INC.5/INF/6, (Nov. 2000), *at* http://www.chem.unep.ch/sc/documents/meetings/inc5/En/inf5-6/inf6e.doc.

All the language in the final text from the fifth INC became the negotiating text for the Diplomatic Conference. In turn, the Diplomatic Conference, while making some changes outlined in Section 8 below, retained the text primarily in the form adopted at INC 5.

I. Conference of Plenipotentiaries on the Stockholm Convention on Persistent Organic Pollutants—Stockholm, 22-23 September 2001

The Government of the Kingdom of Sweden, in Stockholm, convened the Conference of Plenipotentiaries on the Stockholm Convention on Persistent Organic Pollutants May 22-23, 2001.[119] Representatives of 118 countries, nine United Nations bodies and specialized agencies, four intergovernmental organizations and over 100 NGOs attended this Conference.

Before the actual Conference of Plenipotentiaries to adopt the Stockholm Convention, the Secretariat still was faced with some remaining issues from Johannesburg that were note fully resolved. To look into these issues, a preparatory meeting for the conference was called on May 21, and decided on the following issues:[120]

> a) a resolution establishing interim arrangements while the Convention is not yet in force in order to prepare and facilitate the rapid entry into force and effective implementation of the Convention by all signatories, in particular developing countries.[121] For example, promotion of capacity-building programs and assistance on preparation of implementation plans, preparation of reports on POPs inventories, scientific evaluation and alternative to DDT in developing countries in order to prepare for the elimination of DDT, call on States and economic integration organizations to make voluntary contribution in order to support interim activities; and commitment of preparing guidelines, rules of procedure, composition and operational role of the future Persistent Organic Pollutants Review Committee under Article 19, paragraph 6 of the Stockholm Convention;
>
> b) a resolution on liability and redress concerning the use and intentional introduction into the environment of persistent organic pollutants,[122] calling upon the Secretariat to prepare in cooperation with States workshops on liability and redress in the context of the Stockholm Convention and to prepare a report of the workshops to be presented at the next INC meeting with a view to decide what further actions should be taken, which has not yet happened;

---

119 *Final Act of the Conference of Plenipotentiaries on the Stockholm Convention on Persistent Organic Pollutants*, UNEP Doc. UNEP/POPS/CONF/4, (June 2001), *at* http://www.chem.unep.ch/sc/documents/meetings/dipcon/25june2001/conf4_finalact/en/FINALACT-English.doc.

120 *Report of the Preparatory Meeting for the Conference of Plenipotentiaries on the Stockholm Convention on Persistent Organic Pollutants*, UNEP Doc. UNEP/POPS/CONF/PM/3/Rev.1 (May 21, 2001), *at* http://www.chem.unep.ch/sc/documents/meetings/dipcon/25june2001/pm3rev1/k0122164.doc.

121 *See id.* Appendix 1(A) (Resolution on interim arrangements).

122 *See id.* Appendix 1(C)(1)-(3).

c) a resolution on capacity-building and creating a capacity assistance network, requesting INC to provide during the interim period to arrange programs for capacity-building for the implementation of the Stockholm Convention by developing countries and confirming the invitation of GEF as the principal entrusted with the operations of the financial mechanism referred to in Article 13 of the Convention.[123] Between May 2001 and April 11th, 2002, 126 countries signed the Stockholm Convention, including the European Community and five countries have ratified, namely Canada, Fiji, Samoa, Lesotho, and the Netherlands.[124] The Stockholm Convention is still open for signature at the United Nations Headquarters in New York until 22 May 2002.[125]

The Conference of Plenipotentiaries for the Stockholm Convention on Persistent Organic Pollutants proved to be a model of international cooperation to seek solutions for environmental problems. There was great optimism among participants that substantial progress had been achieved. The final structure of the Stockholm Convention is to be evaluated under the following criteria: a) control provisions, b) general provisions, c) procedure for adding new chemical hazardous substances as POPs, and d) financial and technical assistance. The first part, control provisions, contains key core obligations for the elimination of production and use of the initial nine intentionally persistent organic pollutants, listed under Annex A.[126] States that need specific exemption of these above-mentioned POPs will notify the Secretariat, which will keep a public register of all country-specific exemptions.[127] Exemptions will be valid for five years, and they can be renewed upon decision of the Conference of the Parties.

J. The Role of Non-Governmental Organizations in Negotiating the Stockholm Convention

By the beginning of the 1990s, environmental organizations had become increasingly important actors in global environmental politics, organized and accredited as environmental non-governmental organizations at the United Nations. This development united two streams, that of public advocacy long present in domestic life in certain countries such as the U.S. and Britain, and that of large membership groups organized internationally to achieve international peace and understanding. For example, U.S.-based groups in the 19th century, such as the Sierra Club and the several Audubon Societies, reflected an interest in the outdoors and in birding which led to creation of national parks and treaties on migratory birds negotiated on bilateral or regional bases, the

---

123 See id. Appendix 1(B)(1)-(4).
124 *Status of signature and ratification of the Stockholm Convention as of April 11th, 2002*, at http://www.chem.unep.ch/sc/documents/signature/signstatus.htm.
125 See The Stockholm Convention, *supra* note 1, art. 24 (Signature).
126 See The Stockholm Convention, *supra* note 1, Annex A (aldrin, dieldrin, endrin, chlordane, heptachlor, hexachlorobenzene, mirex, toxaphene, and polychlorinated biphenyls).
127 Revised list of requests for specific exemptions in Annex A and B, UNEP Doc. UNEP/POPS/CONF/INF/1/Rev.3, (June 14, 2001), *at* http://www.chem.unep.ch/sc/documents/meetings/dipcon/25june2001/inf1rev3/k0122169.doc.

forerunners of international environmental law. In the mid-twentieth century with the founding of the United Nations, groups known as non-governmental organizations were participants in meetings and were recognized in the UN Charter through ECOSOC.[128] These early NGOs at the UN were primarily large international membership groups such as organizations of women and youth, the World Federation of UN associations, faith-based groups from churches and synagogues, and women's or professional societies, such as the World Council of Churches or the Union des Advocats. One other significant development in environmental protection advocacy on the global scene was the founding of the IUCN in 1948, uniquely composed of representatives of governments and non-profit organizations and scientific and legal experts. By the time of the preparatory conferences to UNCED, an ever-increasing number of non-governmental organizations were routinely attending meetings at the United Nations and influencing the growth of the new field of international environmental law.

Although under the views of traditional public international law, only sovereign states have rights and responsibilities and are members of the United Nations, a more dynamic and modern view of international law recognizes the important contributions of NGOs and their expertise as well as passion, particularly on human rights and environmental law issues.[129] NGOs have played key role in proposing a series of reforms in many fields and at different institutions, such as the International Monetary Fund meetings regarding structural adjustments, World Trade Organization meetings and meetings of the International Whaling Commission in changing the goals from regulation to elimination of whaling NGOs participated in most international negotiations meeting on a wide range of issues leading up to the environmental treaties adopted in the 1990s. Some of the most active NGO participants in the UN on environmental issues include Greenpeace International (founded in the 1970s, now with 3.3 million affiliates), Friends of the Earth (a network of 52 affiliate organization), and the World Wildlife Fund (WWF), a federation of 23 national organizations with over 3 million members.

As the environmental movement has developed internationally, NGOs have formed alliances across national boundaries, increasingly involving colleagues from developing countries. There are networks and coalitions of hundreds of groups such as the Pesticide Action Network, started in Malaysia, which was very active during the POPs negotiations.

The presence of NGO representatives on official delegations to international negotiating committees and multilateral conferences is mutually beneficial, as the government benefits from the expertise and political insight of the NGO repre-

---

128 Charter of the United Nations, Article 71, *at* http://www.un.org/Overview/Charter/chapte10.html (last visited Jan. 29, 2002) ("the Economic and Social Council may make suitable arrangements for consultation with non-governmental organizations which are concerned with matters within its competence. Such arrangements may be made with international organizations and, where appropriate, with national organizations after consultation with the Member of the United Nations concerned.").

129 DAVID HUNTER, JAMES SALZMAN & DURWOOD ZAELKE., INTERNATIONAL ENVIRONMENTAL LAW § IV(A) (Foundation Press, 1998).

sentative, and the NGO representative is able to communicate with other NGOs regarding the transparency of the process.

When decisions are made in isolation, lacking public participation, they are more likely to be ignored or resisted. Accordingly, one of the principles of international environmental law is the participation of NGOs and other elements of civil society.[130] In addition, NGO participation fosters public education. NGO tend to educate the public through the media, by publishing books, information brochures and papers and holding public conferences. It strengthens the democratic process.[131]

Participation of non-governmental organizations to negotiate the Stockholm Convention was elaborated by the First Intergovernmental Negotiation Committee as Rules of Procedure, drawn within the framework of applicable rules of the United Nations. In this respect, the United Nations Economic and Social Council (ECOSOC) adopted Resolution 1996/31,[132] regulating participation of NGOs in international conferences convened by the United Nations and their preparatory process,[133] pursuant to Article 71 of the Charter of the United Nations.[134] NGOs may acquire "consultative status" with ECOSOC and its subsidiary bodies, meaning rights and privileges. This status enables NGOs to make contributions to the work programs and goals of the United Nations, serving as technical experts, advisers and consultants to governments, the Secretariat and UN bodies. Through this resolution, NGOs are allowed to participate in ECOSOC and its various subsidiary bodies through attendance at these meetings, and under some circumstances, with oral interventions and written statements.

In adopting the Rules of Procedures for the meetings of the POPs INCs to negotiate an international legally binding instrument for implementing international action on certain persistent organic pollutants, delegates approved participation of NGOs under the status of observers[135] pursuant United Nations General Assembly, Decision 1/1[136] and 2/1.[137]

---

130 EDITH BROWN WEISS, IN FAIRNESS TO FUTURE GENERATIONS: INTERNATIONAL LAW, COMMON PATRIMONY, AND INTERGENERATIONAL EQUITY (Transnational Publishers 1989).
131 HENKIN PUGH & SCHECHTER SMIT, INTERNATIONAL LAW CASES, CASES AND MATERIAL (3rd ed. West Publishing 1993).
132 Consultative relationship between the United Nations and Non-Governmental Organizations ECOSOC Res. 1996/31, 49th Plenary Meeting (July 25, 1996), *at* http://www.un.org/esa/coordination/ngo/Resolution_1996_31/index.htm.
133 *See id.* Part VII.
134 *See id., supra* note 472.
135 Rules of Procedures for the Intergovernmental Negotiating Committee, UNEP Doc: UNEP/POPS/INC.1/2, Section XII, Rule 55, (July 1998), *at* http://chem.unep.ch/sc/documents/meetings/inc1/inc1-2.htm (Observers).
136 UN/GA Decision 1/1, 45th Sess., (October 17, 1990), *at* http://www.chem.unep.ch/pops/POPs_Inc/INC_1/inf12.htm.
137 UN/GA Decision 2/1, 46th Sess., (May 25, 1991), *at* http://www.chem.unep.ch/sc/documents/meetings/inc1/inf12.htm.

CHAPTER 5

LEGAL ANALYSIS OF THE STOCKHOLM CONVENTION ON PERSISTENT ORGANIC POLLUTANTS

To understand the significant new international law established in the Stockholm Convention requires a article-by-article analysis of this new treaty. For instance, Articles 8 and 13 require close study. Paramount in the entire Convention is the dedication to the scientific basis for the obligations adopted in the treaty,[1] the goal of which is the reduction or elimination of certain chemical substances which affect human health and the environment. Beginning at the first negotiating session, governments such as the U.S. urged that science remain "the guiding principle" as the negotiations progressed, "an open and transparent process that engages the participation of all stakeholders."[2]

New duties under international law have been accepted by states parties to this treaty that agree to:

1. eliminate all nine substances in Annex A, barring submission of exemptions (as, for example, Brazil submitted for heptachlor and chlordane, granting them five additional years to phase out their use and productions);

2. restrict use and production of DDT under Annex B only to the control of mosquitoes causing malaria; and

3. restrict use and production of additional substances under Annex C.

Key to the Convention is the listing of new chemicals in addition to those included in the three annexes. The procedure adopted for adding new chemicals is primarily based on chemistry, as outlined in Annex D of the treaty. Screening criteria are based on: chemical name and identification, proof of persistence, bioaccumulation and potential for long-range environmental transport, and adverse effects for human health and the environment.

Major topics included in the Convention, which were the subject of negotiations during the five INCs described in the previous chapters are:

1) measures to reduce or eliminate releases of POPs into the environment;

2) national implementation plans to be created by each state party and made available to others;

3) information exchange;

4) public information, awareness and education;

---

1 Nicholas A. Robinson, *Legal Systems, Decision Making, and the Science of Earth's Systems: Procedural Missing Links*, 27 Ecology L.Qtly. 1077-1161 (2001).
2 *Report of the First Session of the INC for An International Legally Binding Instrument for Implementing International Action on Certain Persistent Organic Pollutants (POPs)*,15 Earth Negotiations Bull. 1 (July 6, 1998).

5) the powers of the Conference of the Parties and financing mechanisms;

6) dispute resolution; and

7) provisions for amendments to the Convention and Annexes.

The mechanisms for adding new substances to the Annexes and the creation of financial mechanisms were the most hotly debated issues during the INCs. The results adopted in the treaty are discussed below.

Preamble

The political and social underpinnings for the treaty are described in the perambulatory language, which, although non binding, is a basis for interpreting the express commitments. For example, the preamble recognizes the importance of science and the characteristics of POPs, namely their toxic properties, resistance to degradation, bioaccumulation and transport through air and water. Significantly, the preamble begins with the effect of POPs on the natural environment and ecosystems, rather than with human health. Also, the preamble is sensitive to the special effects of POPs on developing countries resulting from local exposure, "in particular impacts upon women and, through them, upon future generations." Human rights and equal rights are noted both in the reference to women, future generations, and indigenous peoples, implying a recognition of blame for the actions of developed countries in carrying out activities which especially impact such groups. The spirit of the negotiators was to take into account the effects of actions on those least able to protect themselves and who themselves were not responsible for causing the harm.

The major principles of international environmental law are referenced in the preamble, including the precautionary principle, sustainable development, intergenerational equity, the polluter pays principle, and the principle of common but differentiated responsibility. The preamble notes that precaution "underlies the concerns of all parties and is embedded" within the Convention.[3] The language of Principle 21 of the Stockholm Declaration, as repeated in the Rio Declaration and the Convention on Biological Diversity, is found verbatim in the preamble to the Stockholm Convention. Transfer of technology is also mentioned, and this convention and other international agreements in the field of trade and environment are mentioned as "mutually supportive."

Another unique aspect of the preamble is the recognition of the Arctic and the effect of POPS on mammals, such as sea lions, whales, polar bears, walruses, and humans, because of the bioaccumulation in their fat tissues from eating contaminated fish and breathing contaminated air and their biomagnification. The preamble praises the principle of common but differentiated responsibilities of states, as contained in Principle 7 of the Rio Declaration on Environment and Development. The preamble recognizes the role of the private sector and NGOs

---

[3] Legal obligations are created in Articles 1 and 8 to observe the precautionary approach. *See also* Annex C; *see* Buccini, *supra* note 355;

in reduction and elimination of discharges, charging the chemical industry to act more ethically toward the planet and urging NGOs to continue their active monitoring in guarding the health of the planet. It appears to be unique in treaty language to "underline" the importance of "manufacturers of persistent organic pollutants taking responsibility for reducing adverse effects caused by their products and for providing information to users, Governments and the public on the hazardous properties of those chemicals." The internalization of environmental costs, found in Principle 16 of the Rio Declaration, and the use of economic instruments based on the "polluter pays" principle are also contained in the preamble, "incorporating the social and environmental costs of a polluting substance into its production processes."[4]

Article 1—Objective

This first Article[5] elects the terminology of the precautionary approach rather than the precautionary principle. The significance of this choice is a weakening of the obligation of precaution, what the European states recognize as a basic principle and what is contained in the Rio Declaration and the preamble to the Biodiversity Convention as "the precautionary principle;" this fundamental principle of international environmental law here is reduced to a mere "approach," a de-emphasis sought by certain major developed states, principally the U.S.[6] However, this does not diminish the intent of the negotiators to prevent harm before it occurs, rather than merely manage the resulting damages after harm occurs.

The precautionary element of the objective is present in the procedure for adding new substances to the lists contained in the annexes to the Convention. Specifically, the Persistent Organic Pollutants Review Committee to be established by the Conference of the Parties may recommend the addition of a substance to the list based on science, as contained in Annex D. Scientific proof of persistence, bio-accumulation, and potential for long-range environmental transport, the three main characteristics of POPs, is required but not scientific certainty as to the harm to humans, in order to list new substances.

One criticism of the language adopted at the Fifth INC as the Objective of the Convention is that it does not call directly for reduction or elimination of POPs. Failing to clearly and unequivocally identify the goals of reduction and

---

4   Elizabeth B. Baldwin, *Reclaiming Our Future: International Efforts to Eliminate the Threat of Persistent Organic Pollutants*, 20 HASTINGS INT'L & COMP. L. REV. 855, fn. 104 (1997), (quoting from WWF Position Paper, World Wide Fund for Nature International, A Framework Convention for the Phase-out and Elimination of POPs 6-7, Appendix B (1996)).

5   As noted *infra* Chapter 2, this important first article was not negotiated until the final INC, just prior to the Diplomatic Conference to adopt the final text. Therefore commentary on the article is not yet published as of the date of this writing. The comments above constitute the candidate's best assessment of the meaning and impact of the article based on his research and experience with the treaty.

6   Catherine Tinker, *State Responsibility and the Precautionary Principle, in* THE PRECAUTIONARY PRINCIPLE AND INTERNATIONAL LAW: THE CHALLENGE OF IMPLEMENTATION 53-71 (David Freestone & Ellen Hey eds., 1996); Interview with Catherine Tinker (Feb. 7, 2002).

elimination of POPs as the objective of the treaty weakens the instrument by substituting only the much vaguer goal of "protecting human health and environment" without specifying the means to do so through the reduction or elimination of POPs.

Article 2—Definitions

Three definitions only are deemed necessary in this treaty. Of greatest interest is the inclusion of "regional economic integration organization" as a potential party to the convention, further defined as "an organization constituted by sovereign States of a given region to which its member States have transferred competence in respect of matters governed by this Convention and which has been duly authorized, in accordance with its internal procedures, to sign, ratify, approve or accede to this Convention." The convention therefore recognizes the trend for states to organize themselves into regional economic blocks such as the European Union, MERCOSUL, and NAFTA.

Article 3—Measures to reduce or eliminate releases from intentional production and use

"Measures to reduce or eliminate releases from intentional production and use" are the first to be dealt with under the treaty and provide the general dedication to "reduce or eliminate" releases. This general language supports the specific language of Annex A and they should be read together in relation to intentional acts. The decision to create Annex A with those chlorinated substances about which consensus exists as to their toxicity, adverse effects, and long-range transport. The nine substances listed in Annex A are designated for elimination of production and use both. In fact, most of these nine substances are already banned in developed countries; the intent of Article 3 and Annex A is to bring developing countries on board in eliminating these same substances.

Article 3 should also be read together with Annex B relating specifically to DDT. A separate annex was created for this substance aimed not at elimination, but only at reduction of use due to its economic and social importance. In line with the recommendation of WHO, Annex B was established for those substances needed for other goals, such as the use of DDT to control malaria-bearing mosquitoes, making it impossible to ban them outright. The compromise was to call for the reduction in both use and production of this substance.

Annex A or Annex B substances cannot be imported or exported except for use permitted (as by exemption), or for "environmentally sound disposal" which means if a developing country does not have the ability to dispose of the chemical, it will be able to export it to a country with the technical capacity to properly dispose of the substance without putting it into the stream of international trade.

The design of the treaty is thus based primarily on chemistry in the initial decision to list a substance on either Annex A or Annex B or Annex C (unintentional production of by-products). Once that decision is made, the rest of the treaty concerns environmental management.

New pesticides or new industrial chemicals which exhibit the characteristics of persistent organic pollutants (as described in Annex D) shall be regulated by states "with the aim of preventing the production and use" of new POPs. This provision circumvents any efforts to avoid regulation by manufacturing new substances which are POPs.

The provision for specific exemptions (for which substances and why) as opposed to general exemptions and for use for an "acceptable purpose" still requires use in a "manner that prevents or minimizes human exposure and release into the environment." The existence of specific exemptions was negotiated in recognition of certain states' needs while still prohibiting general exemptions; for example, Brazil's export of heptachlor-treated lumber, which is a termiticide, is permitted due to Brazil's entry of a specific exemption for this use at the time of signature of the treaty. When the treaty enters into force, upon ratification by 50 states, then Brazil will have a period of 5 years to phase out this special exemption, or to renew the special exemption for another 5-year period.

Article 4—Register of Specific Exemptions

The procedure for registering special exemptions entered by states upon signature is described in this article. States must notify the Secretariat in writing, and the Secretariat shall make the Register available to the public. At the first meeting of the Conference of the Parties, a review process will be established for entries in the Register. (See Article 19 (6)(a) regarding the creation of the Persistent Organic Pollutants Review Committee). A party may withdraw a special exemption at any time.

Article 5—Measures to reduce and eliminate releases from unintentional production

"Measures to reduce and eliminate releases from unintentional production" relate principally to dioxins and furans. Annex C was created to list the substances released by industries as by-products of waste incineration, wood pulp bleaching processes, and other common activities. Annex C also addresses two other substances, chlorophenols and chloranil, released unintentionally during specific chemical production processes, such as textile and leather dyeing and finishing; operation of crematoria; waste oil refineries; and operation of motor vehicles, particularly those burning leaded gasoline.

Domestic environmental legislation will be dramatically improved in many states by compliance with Article 5 (a)(2), particularly the "evaluation of the efficacy of the laws and policies of the Party relating to the management of such releases," by developing an action plan regarding unintentional releases. States must describe existing legislation and methods by which they can improve their compliance with the treaty. A five-year review of the action plan is called for to update the plan and provide a schedule for implementation, including strategies and measures. The treaty is also designed to "promote the application of available, feasible and practical measures that can expeditiously achieve a realistic and meaningful level of release reduction or source elimination." Scientific understanding is at a level that such measures are known and usable, based on "best available techniques," defined in Article 5 (f)(1) as:

> the most effective and advanced stage in the development of activities and their methods of operation which indicate the practical suitability of particular techniques for providing in principle the basis for release limitations designed to prevent and, where that is not practicable, generally to reduce releases of chemicals listed in Annex C and their impact on the environment as a whole.

As defined in Annex C relating to dioxins, PCBs and furans, best available techniques include considerations of likely costs and benefits of a measure as well as consideration of precaution and prevention.

"Best environmental practices" mean the application of the most appropriate combination of environmental control measures and strategies. The original idea of prevention of release was weakened by the final consensus language, which permits considerations of "practicability," undefined in the treaty, to excuse compliance by states parties, especially those that still use waste incineration practices.

Article 6—Measures to reduce or eliminate releases from stockpiles and wastes

Another problem addressed in the treaty is that of so-called obsolete pesticides, those no longer in use but present worldwide in countless gardens and farms. Article 6 requires management of stockpiles and wastes "in a manner protective of human health and the environment" by making an inventory of stockpiles and wastes and categorizing them as Annex A or Annex B (intentionally produced). This management is to be "safe, efficient and environmentally sound."

The export of a stockpile can only be for the purpose of disposal of the substance where the country of origin cannot safely do so to prevent "recovery, recycling, reclamation, direct reuse or alternative uses of persistent organic pollutants." In such cases, prior informed consent procedures will apply, with other relevant international rules. The Conference of the Parties shall cooperate with other treaty bodies, such as that of the Basel Convention on the Transboundary Movement of Hazardous Wastes.

Article 7—Implementation Plans

In a nod to state sovereignty, parties are urged to "develop and *endeavour to implement* a plan for the implementation of its obligations" under the treaty (emphasis added). Direct action, therefore, is left to the discretion of the parties to the treaty.

At the same time, this article recognizes the importance of various stakeholders in the formulation and implementation of this plan, urging consultation with "women's groups and groups involved in the health of children." Each party, therefore, is left to design its own implementation plan, with no international review or approval or minimum standards to ensure compliance with the objectives of the treaty. However, the national implementation plans for POPs are suggested to be integrated into parties' sustainable development strategies, at least "where appropriate." Interesting to note is the possibility that regional economic integration organizations, since they may be parties, will also develop implementation plans if they are signatory to the treaty as the organization, and not only as individual member states. The European Commission has separately signed the treaty.

Article 8—Listing of Chemicals in Annexes A, B and C

This article is key to the convention by establishing a procedure for adding new chemicals to those listed in Annex A, Annex B and Annex C. The ability to add new chemicals to the list directly impacts the chemical industry, that will be reluctant to limit production or use of substances, that may not be imported or exported. Once a state party discovers that a new chemical substance is a POP, notification of the Secretariat is expected, but not required. The language of this article is that "a Party may submit a proposal . . ." Annex D lists criteria, which are 100% based on science for determining whether a new substance has the characteristics of a persistent organic pollutant. However, what a state does with this knowledge is, again, discretionary rather than mandatory. This is a weakness in the provision for listing. Yet the ability to propose new chemicals for listing is significant for the future relevance of the treaty.

Significantly, once a state party does propose the listing of a new chemical substance, the Secretariat forwards it to the Persistent Organic Pollutants Review Committee for consideration under Annex D screening criteria. Members of the Committee include government-designated experts in chemical assessment or management appointed by the COP according to equitable geographical distribution. If the substance passes the screening test, it is then subject to the risk profile information required in Annex E. For example, is the chemical likely to result in long-range environmental transport that would eventually lead to significant adverse effect on human health and/or environmental effects?

In analyzing the risk profile, the Committee cannot say that due to lack of full scientific certainty, the proposal will not be allowed to proceed to final listing of the new substance. The Committee will prepare a "risk management evaluation" including "an analysis of possible control measures for the chemical." The Committee's report with a recommendation will be sent to the Conference of the Parties (COP), who will decide whether to list the substance or not. If the Committee allows the proposal to proceed, the final decision then rests with the COP. If the Committee recommends that the proposal should not proceed, the Committee creates an additional procedure with review and invitation for additional information, meaning scientific data, and reconsideration of the decision. If the Committee again declines to proceed with the proposal, the State Party may challenge this decision, leaving the matter to the next Conference of the Parties to direct the Committee to prepare the risk management evaluation.

Once the Conference of the Parties receives both the risk profile and the risk management evaluation from the Committee, the COP "taking due account of the recommendations of the Committee, including any scientific uncertainty, shall *decide, in a precautionary manner,* whether to list the chemical, and specify its related control measures, in Annexes A, B and/or C" (emphasis added). The treaty clearly uses the content of the precautionary principle while it avoids the use of the phrase "precautionary principle" as such, using other related phrases such as "precautionary manner" here in Article 8 (9) and the actual definition of the precautionary principle itself in Article 8 (7). It may be said, therefore, that the promise in the preamble that "precaution underlies the concerns of all parties" and "is embedded in this Convention" may be realized in fact in this crucial article, which defines the most important procedures created under the treaty.

Article 9—Information Exchange

This article refers to information exchange through a mechanism of communication between parties and the Secretariat, which in turn will inform all other parties. Alternatives to POPs, once discovered, may be communicated directly among parties or through the Secretariat.

Significantly, "information on health and safety of humans and the environment shall not be regarded as confidential." The article continues with a puzzling final sentence stating that "other information" exchanged pursuant to this Convention shall be protected as confidential "as mutually agreed." What information would NOT affect health and safety of humans and the environment if it is a POP? If the amount of a stockpile might be considered confidential, it is difficult to imagine how the basic classification of the substances as a POP could be confidential under any stretch of the imagination, since every step of the procedure in Articles 7 and 8 involves consultation and publication of information.

Article 10—Public Information, Awareness and Education

The previous article concerned information exchange among the parties; this article addresses information for the general public to create awareness and education about POPs. A series of measures are described designed to encourage dissemination of information and public participation in addressing POPs and their health and environmental effects and in developing adequate responses, including input into implementation of the Convention. Training of workers, scientists, educators and technical and managerial personnel is encouraged.

Through workshops, seminars and conferences, NGOs and others are encouraged to develop awareness about POPs, along with other policy and decision makers. Specifically highlighted are parties' goals of development and implementation of educational and public awareness programs at the national and international levels, "especially for women, children and the least educated" on POPs and their health and environmental effects and their alternatives (Article 10 (c)). It is interesting to note that this attention to women appears in the subparagraph linked to public awareness and education together with "children and the least educated," rather than in the subsection referring to "policy and decision-makers." The preamble also links women solely to their reproductive capabilities in recognizing "the health concerns, especially in developing countries, resulting from local exposure to persistent organic pollutants, in particular impacts upon *women and, through them, upon future generations*" (emphasis added). One wonders whether men's impact on future generations is not essential, and whether the impact of dioxins released on non-pregnant women working or living near Ground Zero in New York City is unimportant, or whether rich golfers in the developed countries are equally ignored despite their exposure to pesticides classified as POPs under the treaty. Experience in other UN and multilateral conferences regarding the participation and recognition of major groups, including women; have produced language which was not adopted here in the treaty. For example, the preamble to the Convention on Biological Diversity states:

> Recognizing also the vital role that women play in the conservation and sustainable use of biological diversity and affirming the need for the full participation of women at all levels of policy-making and implementation for biological diversity conservation. . . .

Similarly, it is curious that one of the chief obligations adopted by parties to the treaty is that of maintaining "Pollutant Release and Transfer Registers" including annual quantities of the chemicals listed in Annex A, B or C that are released or disposed of, yet each party is urged merely to "give sympathetic consideration to developing mechanisms for collection and dissemination of this information" (Article 10 (5)).

Article 11—Research, Development and Monitoring

This article encourages research, development and monitoring of POPs, their alternatives, and candidate POPs. In addition to studies of various factors and effects of POPs, socio-economic and cultural impacts and methodologies for making inventories of generating sources and analytical techniques for the measurement of releases are encouraged.

Article 12—Technical Assistance

Developed countries accept the need to honor requests from developing countries and economies in transition for technical assistance to help them meet their obligations under the convention. This article emphasizes the needs of the least developed countries and Small Island developing states.

Article 13—Financial Resources and Mechanisms

This is the second most important article of the convention after Article 8, and it was the last one to be resolved during the negotiations (along with Article 1, "Objectives"). However, the implementation of this article will be delayed until such time as the Conference of the Parties decides to do so. In the meantime, Article 14 on Interim Financial Arrangements, will apply, namely with the Global Environment Facility of the World Bank serving as the institutional entity responsible for the treaty's financial provision. This arrangement could go on indefinitely, as it has with other treaties such as the Biodiversity Convention, leaving the financial administration of funds to aid developing countries to the "restructured" GEF over some time.

When it becomes operational by decision of the COP, however, this article will create a new mechanism for "provision of adequate and sustainable financial resources to developing country parties and parties with economies in transition on a grant or concessional basis to assist in their implementation of the Convention." (Article 13 (6)). Most significantly, contributions to the mechanism "shall be additional to other financial transfers to developing country Parties and Parties with economies in transition."

This article articulates the basis for the common but differentiated responsibility principle by acknowledging the linkage between implementation of the Convention by developing countries and the degree to which developed country Parties provide financial resources, technical assistance and technology transfer, consistent with their responsibility for the production and use of POPs. In addition, the treaty recognizes that:

> The fact that sustainable economic and social development and eradication of poverty are the first and overriding priorities of the developing country Parties will be taken fully into account, giving due consideration to the need for the protection of human health and the environment.

The delay in negotiating this language had to do primarily with the subject being money, and the acceptance by developed countries of the obligation to provide funds for developing countries as outlined above.

Article 14—Interim Financial Arrangements

The acceptance of the existing mechanism of the Global Environment Facility, already used by the Convention on Biological Diversity and other treaties, as the interim financial mechanism is detailed here. The importance of this election, until such time as it is changed by the COP at or after their first meeting, is that the flow of funds may commence immediately for capacity building, enabling developing countries to meet standards for future implementation of the treaty. By the time the treaty enters into force, therefore, developing countries may have their inventories ready and technical personnel trained.

Article 15—Reporting

Reporting requirements are established for parties to periodically inform the Secretariat of their compliance through provision of statistical data on its total quantities of production, import and export of each of the chemicals listed in Annex A and Annex B. This obligation applies to both developed and developing countries. The aim is to provide data on who produces POPs and how much, and where does it go.

Article 16—Effectiveness Evaluation

Effectiveness evaluation shall be performed every four years. Arrangements shall be made at the first Conference of the Parties for monitoring data on the presence of the listed chemicals in Annexes A, B and C and on their regional and global environmental transport. This self-reporting mechanism will assist in keeping the convention functional and moving forward, and allow for flexibility in accommodating new scientific knowledge and technical developments.

Article 17—Non-compliance

Non-compliance, according to this article, carries no penalty, a phenomenal statement in a treaty, which appears to create obligations, at least to some extent, of "reducing and eliminating use and production of POPs." The Conference of the Parties shall, "as soon as practicable, develop and approve procedures and institutional mechanisms" both for determining non-compliance and treatment of parties found to be in non-compliance. The normal treaty rules in the Vienna Convention on the Law of Treaties regarding breach and the consequences of breach of treaty obligations would, one would expect, apply. But does this article preclude such remedies? If states become signatory to the convention without knowing the consequences of failure to comply with treaty duties, what will happen when the COP determines treatment of parties in non-compliance some day? Will states be willing to ratify such an open-ended document? Perhaps this article appears in this form because there is no clear legal duty to do anything under the Convention; therefore non-compliance cannot be easily determined or defined. For example, the Article 1 language on "Objective" is vague and general: to protect human health and the environment from persistent organic pollutants." How this is to be done seems to be a function of the listing of certain substances in Annex A (pesticides), Annex B (DDT), or Annex C (unintentional production of POPs as byproducts), with different obligations for reduction or elimination of production and use.

Article 18—Settlement of Disputes

The only mandatory form of dispute resolution required under this article is negotiation. A party may declare in a written instrument that it recognizes either or both arbitration and submission of a dispute to the International Court of Justice as compulsory for any disputes concerning the interpretation or application of the Convention. This procedure is common in other environmental treaties, such as the Climate Change Convention and the Biodiversity Convention. Note that regional economic integration organization parties can demand arbitration as compulsory, but cannot appear before the International Court of Justice, where only states are recognized.

An additional mechanism is suggested, that of creation of a conciliation commission, procedures for which will be established in another annex at some future date by the Conference of the Parties. In at least one other treaty that referred to conciliation, the procedures for both arbitration and conciliation were established in an annex as part of the text of the treaty prior to opening the instrument for signature (Convention on Biological Diversity).

Chapter 5

OPERATIONAL CONSIDERATIONS OF THE CONVENTION

Article 19—Conference of the Parties

Within one year of entry into force of the Convention, the Executive Director of UNEP is directed to convene the first Conference of the Parties, where procedures and financial rules for itself, any subsidiary bodies, and the Secretariat, will be adopted. An important task at the first COP will be the creation of the Persistent Organic Pollutants Review Committee, described above in connection with Article 8. Voting will be by consensus, or, if that fails, by a two-thirds vote of those present and voting.

Article 20 –Secretariat

The Secretariat for the Convention will be the United Nations Environment Programme, unless the COP changes this by a three-fourths majority vote.

Article 21—Amendments to the Convention

Amendments may be offered with six months' notice prior to the next COP. Amendments once ratified, accepted, or approved, shall be notified to the depositary in writing.

Article 22—Adoption and Amendments of Annexes

Interestingly, this article states that "any additional annexes shall be restricted to procedural, scientific, technical or administrative matters." It is unclear what else remains. Written notice of non-acceptance of an annex may be provided by a party to the depositary. Slightly different procedures are established for amending Annex A, B and C as opposed to Annex D, E or F.

Article 23—Right to Vote

The right to vote is on the basis of one party, one vote. The number of votes is equal to the number of its member states that are parties to this convention. But if any member state votes, then the regional organization of which it is a party cannot, and vice versa.

Article 24—Signature

The Convention was opened for signature on 24 May 2001 in Stockholm, Sweden. As of March 25th, 2003, there are 151 signatories, including one regional economic integration organization (the EU). It is open for signature until May 22, 2002, at United Nations headquarters in New York. The signatories[7] include all Nordic states, Canada and the United States, most European Union states, China, Brazil, Ghana, Cameroon, South Africa, Australia, New Zealand, Georgia, Former Yugoslav Republic of Macedonia, Fiji, Philippines representing the major

---

7 Status of Signature of the Stockholm Convention on Persistent Organic Pollutants, *at* http://www.pops.int/documents/signature/signstatus.htm (This reflects the status of signature as of March 25th, 2003).

industrialized states as well as small island states, developing countries and countries with economies in transition. A major non-signatory to date is Russia.

Article 25—Ratification, Acceptance, Approval or Accession

Thirty states have ratified the Stockholm Convention as of March 25th, 2003: Canada, Fiji, Lesotho, Samoa, the Netherlands, and Germany.[8] Sweden, one of the principal states active in the Intergovernmental Negotiating Committees, has not yet ratified, although it signed, the treaty.

Regional economic integration organizations may become parties to the convention without any member state becoming a party, and the member states will be bound. If one or more member state is a party to the Convention, the organization and its member states shall decide upon their respective responsibilities for the performance of their obligations under the treaty. Their rights cannot be exercised concurrently.

Article 26—Entry into Force

The Convention will enter into force on the ninetieth day after the date of deposit of the fiftieth instrument of ratification, acceptance, approval or accession.

Article 27—Reservations

No reservations may be made to this Convention. A similar provision was included in other environmental treaties, such as the Convention on Biological Diversity.

Article 28—Withdrawal

Three years after the Convention has entered into force for a given party, that party may withdraw upon written notification to the depositary. Withdrawal takes place one year from receipt of notification or as specified in the notice.

Article 29—Depositary

The Secretary-General of the United Nations shall be the depositary of this Convention.

Article 30—Authentic Texts

The authentic text of the Convention shall be in all six languages of the United Nations equally.

---

8   *See id.* (Ratification Status as of March 25th, 2003).

CONCLUSION

The subject of persistent organic pollutants has taken center stage in the world's consciousness due to the serious harm caused to human health and the environment from the release of these substances and due to their very characteristics of long-range movement, bioaccumulation, and persistence. In other words, these substances will be with us in the atmosphere, the lithosphere and the hydrosphere long into the foreseeable future and even beyond. In humans, these substances accumulate in the fatty tissue, posing particularly serious health risks such as cancer and endocrine disruption.

Certain domestic activities, such as spraying of pesticides, operating chemical factories, improper discard of chemicals containers, etc, threaten human health and the environment. Such activities occur within the jurisdiction of one sovereign state, but the effects cross borders to create severe hazards to human health and the environment in other countries. This is particularly true in the case of POPs, in which the pollution originates in one state but causes health and environmental damage in others due to transboundary air and water movements of POPs. New methods of agriculture and forestry proved to have unexpected and undesirable hazards. In 1962, Rachel Carson warned us of a spring that was falling silent as the birds disappeared, poisoned by the chemical substances used by farmers to control pests and fungal diseases.[1] Another field of concern is the degradation of our commons that is the environmental resources that do not fall within any state's jurisdiction, such as our oceans and air. Soon we learned through science that pollution was something that did not just affect bodies of waters and fauna in the fields, but had entered our homes in the form of pesticide residue in fruits and vegetables posing a threat to human health. Therefore, hazards posed by POPs are multilateral, meaning that all adverse effects are being produced by the release of POPs from many countries. The reduction in the amounts of POPs from one state will not, as a general matter, materially mitigate the threat posed by these toxic substances. All measures have to be multilateral, and this bringing us to the issue that countries will have to regulate their domestic activities in a harmonized way in order to prevent degradation of common environmental resources. In situations such as climate change, ozone depletion, and POPs, States cannot unilaterally, bilaterally, or regionally make much of a dent in the global problem. This, of course, causes the solution to be much more complex and difficult. In achieving this objective countries will have to deal with political, economical, and legal issues in order to comply with International Environmental Law.

Political and economic considerations complicate addressing immediate elimination of production or use of persistent organic pollutants in some states. Especially in developing countries, a different perspective in environmental conservation and prevention is a reality. Most developing nations' priority is eco-

---

1   Rachel Carson, Silent Spring, (Houghton Mifflin, 1962).

nomic development, and they see environmental protection as a luxury. The other major issue is that developing countries are skeptical that industrialized nations will provide the financial and technical assistance needed to make conservation an equitable burden. However, developing countries cannot continue to believe that developed countries will grant them a better deal to get them into the treaty regime. POPs is a problem of environmental responsibility; as such, developed nations will have to set the example and bear the burden initially since a great deal of the production and release of POPs is created by their generation of pollution and the developing countries cannot act without their financial and technical help.

From a legal standpoint, another distinguishing characteristic of POPs is that they appear to have the greatest effect on populations and regions far from the location of their production. For example, POPs are transported by air and water currents to the North Pole, where the Inuits are suffering from incidents of POPs poisoning in a larger percentage than most other humans elsewhere in the globe despite the fact that they do not use or produce chemical substances containing POPs. Also causing concern is the fact that infants and children who are breastfed by mothers contaminated with POPs also show contamination. Fauna at the North Pole have a great amount of fatty tissue as protection against the extreme cold climate, rendering them more susceptible to bioaccumulation of POPs, which are fat-soluble rather than water-soluble, than animals in other climates. The contamination is passed along the food chain.

The Stockholm Convention will star the process of protecting the biosphere and human health because it is the first international legal instrument to require the reduction and elimination of persistent organic pollutants, rather than attempting to control their trade or movement as other treaties have done for hazardous wastes or other substances (Basel Convention), or establishing procedures for prior informed consent before the international movement of chemicals (Rotterdam Convention). The treaty itself acknowledges that the only solution to the problem of POPs is the eventual elimination of their production and use, as in many cases there are no other known means of controlling the substances once manufactured or applied to fields or bodies of water. There are no economically feasible means of remediation or clean-up of POPs, such as PCBs collected from old transformers, since they are not easily stored or decontaminated. Some POPs are common pesticides such as DDT and chlordane; some are results of industrial production such as dioxins and furans (principally from waste incineration); and others are industrial chemicals such as PCBs.

It is difficult to state how long it will be before the effectiveness of the Stockholm Convention can be judged. The treaty was opened for signature on May 24, 2001, and will close for signatures on May 22, 2002. As of March 25th, 2003, 151 states have signed the treaty and of those, thirty states have ratified the convention. Fifty states must ratify before the convention enters into force. Domestic legal and political considerations may delay ratification even in states most active in the negotiation of the treaty, such as Sweden. Other leaders in the treaty negotiations, such as Brazil, have notoriously slow processes for ratification of international treaties. Canada, on the other hand, one of the first to ratify, wanted to set the example and continue its leadership in international environmental law while protecting the most vulnerable population in the world, the Inuits. The Netherlands, Fiji, Lesotho, and Samoa are the others who have ratified as of April 11th, 2002. The last country to ratify the Convention was Panama on March 5th, 2003.

In India, the domestic chemical industry is opposed to the treaty because they are major manufacturers of DDT, which is also used extensively in India and exported elsewhere. The U.S. will have to change domestic legislation to comply with the treaty (FIFRA, TSCA, and possibly the Clean Water Act and Clean Air Act as well), requiring research and a political process to accomplish this end. The EU has signed the treaty, and has a clear directive for control and management of chemical substances within the European Union. Each member state of the EU will have to ratify the treaty separately, although only the Netherlands has done so to date. By the end of 2002, most European states are expected to have ratified the treaty.

The shortcomings of the treaty are few but may be significant, namely the language adopted in the final text after extended battles from various sides and interests is weak. The treaty commits its states parties to "protect human health and the environment" without any more specific language in its objective. Specific language was rejected in the end which would have required all states parties to the treaty to *eliminate* all twelve substances listed in the treaty. Thus the primary legal obligation is vague and non-specific and offers little or no guidance to states wishing to comply. In addition, enforcement of the treaty's objective will be difficult since there are no concrete steps or actions required which can be monitored and evaluated for compliance. In seeking consensus, and attempting to adopt an international legal instrument based on clear science, the legal language is quite weak and innocuous, or so broad as to be almost meaningless in terms of enforcement. In addition, the treaty does not contain any sanctions or consequences for non-compliance, leaving this subject to subsequent debate at the Conference of the Parties, a standard technique in international environmental law-making.

How, then shall we measure the success of the Stockholm Convention? One way is to see its very existence as a success for the international community and the global commons. Another measure of its success will be the addition of new chemical substances to the annexes to the convention, bringing more chemicals under restrictions in the name of protecting human health and the environment. Effective implementation of the provisions for public information, awareness and education will lead to another measure of success as the general public demands more protection from POPs and policy-makers alter their decisions accordingly. Ultimately, another mark of success will be when developing countries are helped by others through sharing of costs to assess risks more effectively and find alternatives to POPs for important policy goals such as disease control and agricultural needs. The treaty provides a mechanism for cost-sharing, which, if utilized both by donors and by recipients, will be another sign of success of the treaty. Finally, an important measure of the success of the Stockholm Convention will be the strengthening of domestic municipal legal systems, data collection, information exchange, and capabilities for management of chemicals in order to better protect and inform individuals of the risks of POPs.

The legal innovations in the treaty are few. The annex device for listing substances has been used before; the technique of putting principles of customary international law into the preamble and the choice to draft a vague, general objective have all been used before in international environmental treaties. Yet the many references to precaution and the very concept of the treaty itself may actually advance the acceptance of the precautionary principle beyond other treaties. For example, even those states which insist on a "precautionary approach" rather than a binding "precautionary principle" in this subject area have adopted a total ban on POPs within their own states. Accordingly, the approach accomplishes the same thing as a principle for this purpose. The Stockholm Convention may be seen as a contribution to and an example of a forceful means of putting the precautionary principle into action. The real innovation of the treaty is the adoption of a full annex, Annex F, on socio-economic considerations:

Annex F

... relevant information should be provided relating to socio-economic considerations associated with possible control measures.... Such information should reflect due regard for the differing capabilities and conditions among the Parties and should include consideration of...

(a) Efficacy and efficiency of possible control measures in meeting risk reduction goals: (i) Technical feasibility; and (ii) Costs, including environmental and health costs;

(b) Alternatives (products and processes): (i) Technical feasibility; (ii) Costs, including environmental and health costs; (iii) Efficacy; (iv) Risk; (v) Availability; and (vi) Accessibility;

(c) Positive and/or negative impacts on society of implementing possible control measures: (i) Health, including public, environmental and occupational health; (ii) Agriculture, including aquaculture and forestry; (iii) Biota (biodiversity); (iv) Economic aspects; (v) Movement towards sustainable development; and (vi) Social costs;

(d) Waste and disposal implications (in particular, obsolete stocks of pesticides and clean-up of contaminated sites): (i) Technical feasibility; and (ii) Cost;

(e) Access to information and public education;

(f) Status of control and monitoring capacity; and

(g) Any national or regional control actions taken, including information on alternatives, and other relevant risk management information.

In the framework of environmental law, the Stockholm Convention's significance is notable as the first international agreement on the subject of chemical substances to address human health and protection of the environment. This newest international environmental treaty is groundbreaking in its subject matter and will help developing nations by financing their internal systems for risk assessment, chemical inventories, and adoption of less-harmful alternatives while achieving important public health, industrial, agricultural, and other priorities. Developed countries too will benefit, since their citizens are currently at risk of POPs contamination from foodstuffs imported back into developed states from the developing states where they are routinely used for agriculture. The additional burden to developed countries is small, in that they have already prohibited the use of these substances in their own countries, by and large. Now it will be up to the developing world to take the same steps, with due regard for socio-economic considerations.

APPENDIX

TEXT OF THE STOCKHOLM CONVENTION ON PERSISTENT ORGANIC POLLUTANTS

The Parties to this Convention,

Recognizing that are transported, through air, water and migratory species, across international boundaries and deposited far from their place of release, where they accumulate in terrestrial and aquatic ecosystems,

Aware of the health concerns, especially in developing countries, resulting from local exposure to persistent organic pollutants, in particular impacts upon women and, through them, upon future generations,

Acknowledging that the Arctic ecosystems and indigenous communities are particularly at risk because of the biomagnification of persistent organic pollutants and that contamination of their traditional foods is a public health issue,

Conscious of the need for global action on persistent organic pollutants,

Mindful of decision 19/13 C of 7 February 1997 of the Governing Council of the United Nations Environment Programme to initiate international action to protect human health and the environment through measures which will reduce and/or eliminate emissions and discharges of persistent organic pollutants,

Recalling the pertinent provisions of the relevant international environmental conventions, especially the Rotterdam Convention on the Prior Informed Consent Procedure for Certain Hazardous Chemicals and Pesticides in International Trade, and the Basel Convention on the Control of Transboundary Movements of Hazardous Wastes and their Disposal including the regional agreements developed within the framework of its Article 11,

Recalling also the pertinent provisions of the Rio Declaration on Environment and Development and Agenda 21,

Acknowledging that precaution underlies the concerns of all the Parties and is embedded within this Convention,

Recognizing that this Convention and other international agreements in the field of trade and the environment are mutually supportive,

Reaffirming that States have, in accordance with the Charter of the United Nations and the principles of international law, the sovereign right to exploit their own resources pursuant to their own environmental and developmental policies, and the responsibility to ensure that activities within their jurisdiction or control do not cause damage to the environment of other States or of areas beyond the limits of national jurisdiction,

Taking into account the circumstances and particular requirements of developing countries, in particular the least developed among them, and countries with economies in transition, especially the need to strengthen their national capabilities for the management of chemicals, including through the transfer of technology, the provision of financial and technical assistance and the promotion of cooperation among the Parties,

Taking full account of the Programme of Action for the Sustainable Development of Small Island Developing States, adopted in Barbados on 6 May 1994,

Noting the respective capabilities of developed and developing countries, as well as the common but differentiated responsibilities of States as set forth in Principle 7 of the Rio Declaration on Environment and Development,

Recognizing the important contribution that the private sector and non-governmental organizations can make to achieving the reduction and/or elimination of emissions and discharges of persistent organic pollutants,

Underlining the importance of manufacturers of persistent organic pollutants taking responsibility for reducing adverse effects caused by their products and for providing information to users, Governments and the public on the hazardous properties of those chemicals,

Conscious of the need to take measures to prevent adverse effects caused by persistent organic pollutants at all stages of their life cycle,

Reaffirming Principle 16 of the Rio Declaration on Environment and Development which states that national authorities should endeavour to promote the internalization of environmental costs and the use of economic instruments, taking into account the approach that the polluter should, in principle, bear the cost of pollution, with due regard to the public interest and without distorting international trade and investment,

Encouraging Parties not having regulatory and assessment schemes for pesticides and industrial chemicals to develop such schemes,

Recognizing the importance of developing and using environmentally sound alternative processes and chemicals,

Determined to protect human health and the environment from the harmful impacts of persistent organic pollutants,

Have agreed as follows:

Article 1
Objective

Mindful of the precautionary approach as set forth in Principle 15 of the Rio Declaration on Environment and Development, the objective of this Convention is to protect human health and the environment from persistent organic pollutants.

Article 2
Definitions

For the purposes of this Convention:

(a) "Party" means a State or regional economic integration organization that has consented to be bound by this Convention and for which the Convention is in force;

(b) "Regional economic integration organization" means an organization constituted by sovereign States of a given region to which its member States have transferred competence in respect of matters governed by this Convention and which has been duly authorized, in accordance with its internal procedures, to sign, ratify, accept, approve or accede to this Convention;

(c) "Parties present and voting" means Parties present and casting an affirmative or negative vote.

Article 3
Measures to reduce or eliminate releases from intentional production and use

1. Each Party shall:

   (a) Prohibit and/or take the legal and administrative measures necessary to eliminate:

   (i) Its production and use of the chemicals listed in Annex A subject to the provisions of that Annex; and

   (ii) Its import and export of the chemicals listed in Annex A in accordance with the provisions of paragraph 2; and

   (b) Restrict its production and use of the chemicals listed in Annex B in accordance with the provisions of that Annex.

2. Each Party shall take measures to ensure:

   (a) That a chemical listed in Annex A or Annex B is imported only:

   (i) For the purpose of environmentally sound disposal as set forth in paragraph 1 (d) of Article 6; or

   (ii) For a use or purpose which is permitted for that Party under Annex A or Annex B;

   (b) That a chemical listed in Annex A for which any production or use specific exemption is in effect or a chemical listed in Annex B for which any production or use specific exemption or acceptable purpose is in effect, taking into account any relevant provisions in existing international prior informed consent instruments, is exported only:

   (i) For the purpose of environmentally sound disposal as set forth in paragraph 1 (d) of Article 6;

(ii) To a Party which is permitted to use that chemical under Annex A or Annex B; or

(iii) To a State not Party to this Convention which has provided an annual certification to the exporting Party. Such certification shall specify the intended use of the chemical and include a statement that, with respect to that chemical, the importing State is committed to:

a. Protect human health and the environment by taking the necessary measures to minimize or prevent releases;

b. Comply with the provisions of paragraph 1 of Article 6; and

c. Comply, where appropriate, with the provisions of paragraph 2 of Part II of Annex B.

The certification shall also include any appropriate supporting documentation, such as legislation, regulatory instruments, or administrative or policy guidelines. The exporting Party shall transmit the certification to the Secretariat within sixty days of receipt.

(c) That a chemical listed in Annex A, for which production and use specific exemptions are no longer in effect for any Party, is not exported from it except for the purpose of environmentally sound disposal as set forth in paragraph 1 (d) of Article 6;

(d) For the purposes of this paragraph, the term "State not Party to this Convention" shall include, with respect to a particular chemical, a State or regional economic integration organization that has not agreed to be bound by the Convention with respect to that chemical.

3. Each Party that has one or more regulatory and assessment schemes for new pesticides or new industrial chemicals shall take measures to regulate with the aim of preventing the production and use of new pesticides or new industrial chemicals which, taking into consideration the criteria in paragraph 1 of Annex D, exhibit the characteristics of persistent organic pollutants.

4. Each Party that has one or more regulatory and assessment schemes for pesticides or industrial chemicals shall, where appropriate, take into consideration within these schemes the criteria in paragraph 1 of Annex D when conducting assessments of pesticides or industrial chemicals currently in use.

5. Except as otherwise provided in this Convention, paragraphs 1 and 2 shall not apply to quantities of a chemical to be used for laboratory-scale research or as a reference standard.

6. Any Party that has a specific exemption in accordance with Annex A or a specific exemption or an acceptable purpose in accordance with Annex B shall take appropriate measures to ensure that any production or use under such exemption or purpose is carried out in a manner that prevents or minimizes human exposure and release into the environment. For exempted uses or acceptable purposes that involve intentional release into the environment under conditions

of normal use, such release shall be to the minimum extent necessary, taking into account any applicable standards and guidelines.

Article 4
Register of specific exemptions

1. A Register is hereby established for the purpose of identifying the Parties that have specific exemptions listed in Annex A or Annex B. It shall not identify Parties that make use of the provisions in Annex A or Annex B that may be exercised by all Parties. The Register shall be maintained by the Secretariat and shall be available to the public.

2. The Register shall include:

   (a) A list of the types of specific exemptions reproduced from Annex A and Annex B;

   (b) A list of the Parties that have a specific exemption listed under Annex A or Annex B; and

   (c) A list of the expiry dates for each registered specific exemption.

3. Any State may, on becoming a Party, by means of a notification in writing to the Secretariat, register for one or more types of specific exemptions listed in Annex A or Annex B.

4. Unless an earlier date is indicated in the Register by a Party, or an extension is granted pursuant to paragraph 7, all registrations of specific exemptions shall expire five years after the date of entry into force of this Convention with respect to a particular chemical.

5. At its first meeting, the Conference of the Parties shall decide upon its review process for the entries in the Register.

6. Prior to a review of an entry in the Register, the Party concerned shall submit a report to the Secretariat justifying its continuing need for registration of that exemption. The report shall be circulated by the Secretariat to all Parties. The review of a registration shall be carried out on the basis of all available information. Thereupon, the Conference of the Parties may make such recommendations to the Party concerned as it deems appropriate.

7. The Conference of the Parties may, upon request from the Party concerned, decide to extend the expiry date of a specific exemption for a period of up to five years. In making its decision, the Conference of the Parties shall take due account of the special circumstances of the developing country Parties and Parties with economies in transition.

8. A Party may, at any time, withdraw an entry from the Register for a specific exemption upon written notification to the Secretariat. The withdrawal shall take effect on the date specified in the notification.

9. When there are no longer any Parties registered for a particular type of specific exemption, no new registrations may be made with respect to it.

Article 5
Measures to reduce or eliminate releases from unintentional production

Each Party shall at a minimum take the following measures to reduce the total releases derived from anthropogenic sources of each of the chemicals listed in Annex C, with the goal of their continuing minimization and, where feasible, ultimate elimination:

(a) Develop an action plan or, where appropriate, a regional or subregional action plan within two years of the date of entry into force of this Convention for it, and subsequently implement it as part of its implementation plan specified in Article 7, designed to identify, characterize and address the release of the chemicals listed in Annex C and to facilitate implementation of subparagraphs (b) to (e). The action plan shall include the following elements:

(i) An evaluation of current and projected releases, including the development and maintenance of source inventories and release estimates, taking into consideration the source categories identified in Annex C;

(ii) An evaluation of the efficacy of the laws and policies of the Party relating to the management of such releases;

(iii) Strategies to meet the obligations of this paragraph, taking into account the evaluations in (i) and (ii);

(iv) Steps to promote education and training with regard to, and awareness of, those strategies;

(v) A review every five years of those strategies and of their success in meeting the obligations of this paragraph; such reviews shall be included in reports submitted pursuant to Article 15;

(vi) A schedule for implementation of the action plan, including for the strategies and measures identified therein;

(b) Promote the application of available, feasible and practical measures that can expeditiously achieve a realistic and meaningful level of release reduction or source elimination;

(c) Promote the development and, where it deems appropriate, require the use of substitute or modified materials, products and processes to prevent the formation and release of the chemicals listed in Annex C, taking into consideration the general guidance on prevention and release reduction measures in Annex C and guidelines to be adopted by decision of the Conference of the Parties;

(d) Promote and, in accordance with the implementation schedule of its action plan, require the use of best available techniques for new sources within source categories which a Party has identified as warranting such action in its action plan, with a particular initial focus on source categories identified in Part II of Annex C. In any case, the requirement to use best available techniques for new sources in the categories listed in Part II of that Annex shall be phased in as soon as practicable but no later than four years after the entry into force of the Convention for that Party. For the identified categories, Parties shall promote the use of best environmental practices. When applying best available techniques and best environmental practices, Parties should take into consideration the general guidance on prevention and release reduction measures in that Annex and guidelines on best available techniques and best environmental practices to be adopted by decision of the Conference of the Parties;

(e) Promote, in accordance with its action plan, the use of best available techniques and best environmental practices:

(i) For existing sources, within the source categories listed in Part II of Annex C and within source categories such as those in Part III of that Annex; and

(ii) For new sources, within source categories such as those listed in Part III of Annex C which a Party has not addressed under subparagraph (d).

When applying best available techniques and best environmental practices, Parties should take into consideration the general guidance on prevention and release reduction measures in Annex C and guidelines on best available techniques and best environmental practices to be adopted by decision of the Conference of the Parties;

(f) For the purposes of this paragraph and Annex C:

(i) "Best available techniques" means the most effective and advanced stage in the development of activities and their methods of operation which indicate the practical suitability of particular techniques for providing in principle the basis for release limitations designed to prevent and, where that is not practicable, generally to reduce releases of chemicals listed in Part I of Annex C and their impact on the environment as a whole.

In this regard:

(ii) "Techniques" includes both the technology used and the way in which the installation is designed, built, maintained, operated and decommissioned;

(iii) "Available" techniques means those techniques that are accessible to the operator and that are developed on a scale that allows implementation in the relevant industrial sector, under economically and technically viable conditions, taking into consideration the costs and advantages; and

(iv) "Best" means most effective in achieving a high general level of protection of the environment as a whole;

(v) "Best environmental practices" means the application of the most appropriate combination of environmental control measures and strategies;

(vi) "New source" means any source of which the construction or substantial modification is commenced at least one year after the date of:

a. Entry into force of this Convention for the Party concerned; or

b. Entry into force for the Party concerned of an amendment to Annex C where the source becomes subject to the provisions of this Convention only by virtue of that amendment.

(g) Release limit values or performance standards may be used by a Party to fulfill its commitments for best available techniques under this paragraph.

Article 6
Measures to reduce or eliminate releases from stockpiles and wastes

1. In order to ensure that stockpiles consisting of or containing chemicals listed either in Annex A or Annex B and wastes, including products and articles upon becoming wastes, consisting of, containing or contaminated with a chemical listed in Annex A, B or C, are managed in a manner protective of human health and the environment, each Party shall:

(a) Develop appropriate strategies for identifying:

(i) Stockpiles consisting of or containing chemicals listed either in Annex A or Annex B; and

(ii) Products and articles in use and wastes consisting of, containing or contaminated with a chemical listed in Annex A, B or C;

(b) Identify, to the extent practicable, stockpiles consisting of or containing chemicals listed either in Annex A or Annex B on the basis of the strategies referred to in subparagraph (a);

(c) Manage stockpiles, as appropriate, in a safe, efficient and environmentally sound manner. Stockpiles of chemicals listed either in Annex A or Annex B, after they are no longer allowed to be used according to any specific exemption specified in Annex A or any specific exemption or acceptable purpose specified in Annex B, except stockpiles which are allowed to be exported according to paragraph 2 of Article 3, shall be deemed to be waste and shall be managed in accordance with subparagraph (d);

(d) Take appropriate measures so that such wastes, including products and articles upon becoming wastes, are:

(i) Handled, collected, transported and stored in an environmentally sound manner;

(ii) Disposed of in such a way that the persistent organic pollutant content is destroyed or irreversibly transformed so that they do not exhibit the characteristics of persistent organic pollutants or otherwise disposed of in an environmentally sound manner when destruction or irreversible transformation does not represent the environmentally preferable option or the persistent organic pollutant content is low, taking into account international rules, standards, and guidelines, including those that may be developed pursuant to paragraph 2, and relevant global and regional regimes governing the management of hazardous wastes;

(iii) Not permitted to be subjected to disposal operations that may lead to recovery, recycling, reclamation, direct reuse or alternative uses of persistent organic pollutants; and

(iv) Not transported across international boundaries without taking into account relevant international rules, standards and guidelines;

(e) Endeavour to develop appropriate strategies for identifying sites contaminated by chemicals listed in Annex A, B or C; if remediation of those sites is undertaken it shall be performed in an environmentally sound manner.

2. The Conference of the Parties shall cooperate closely with the appropriate bodies of the Basel Convention on the Control of Transboundary Movements of Hazardous Wastes and their Disposal to, *inter alia*:

(a) Establish levels of destruction and irreversible transformation necessary to ensure that the characteristics of persistent organic pollutants as specified in paragraph 1 of Annex D are not exhibited;

(b) Determine what they consider to be the methods that constitute environmentally sound disposal referred to above; and

(c) Work to establish, as appropriate, the concentration levels of the chemicals listed in Annexes A, B and C in order to define the low persistent organic pollutant content referred to in paragraph 1 (d)(ii).

Article 7
Implementation plans

1. Each Party shall:

(a) Develop and endeavour to implement a plan for the implementation of its obligations under this Convention;

(b) Transmit its implementation plan to the Conference of the Parties within two years of the date on which this Convention enters into force for it; and

(c) Review and update, as appropriate, its implementation plan on a periodic basis and in a manner to be specified by a decision of the Conference of the Parties.

2. The Parties shall, where appropriate, cooperate directly or through global, regional and subregional organizations, and consult their national stakeholders, including women's groups and groups involved in the health of children, in order to facilitate the development, implementation and updating of their implementation plans.

3. The Parties shall endeavour to utilize and, where necessary, establish the means to integrate national implementation plans for persistent organic pollutants in their sustainable development strategies where appropriate.

Article 8
Listing of chemicals in Annexes A, B and C

1. A Party may submit a proposal to the Secretariat for listing a chemical in Annexes A, B and/or C. The proposal shall contain the information specified in Annex D. In developing a proposal, a Party may be assisted by other Parties and/or by the Secretariat.

2. The Secretariat shall verify whether the proposal contains the information specified in Annex D. If the Secretariat is satisfied that the proposal contains the information so specified, it shall forward the proposal to the Persistent Organic Pollutants Review Committee.

3. The Committee shall examine the proposal and apply the screening criteria specified in Annex D in a flexible and transparent way, taking all information provided into account in an integrative and balanced manner.

4. If the Committee decides that:

   (a) It is satisfied that the screening criteria have been fulfilled, it shall, through the Secretariat, make the proposal and the evaluation of the Committee available to all Parties and observers and invite them to submit the information specified in Annex E; or

   (b) It is not satisfied that the screening criteria have been fulfilled, it shall, through the Secretariat, inform all Parties and observers and make the proposal and the evaluation of the Committee available to all Parties and the proposal shall be set aside.

5. Any Party may resubmit a proposal to the Committee that has been set aside by the Committee pursuant to paragraph 4. The resubmission may include any concerns of the Party as well as a justification for additional consideration by the Committee. If, following this procedure, the Committee again sets the proposal aside, the Party may challenge the decision of the Committee and the Conference of the Parties shall consider the matter at its next session. The Conference of the Parties may decide, based on the screening criteria in Annex D and taking into account the evaluation of the Committee and any additional information provided by any Party or observer, that the proposal should proceed.

6. Where the Committee has decided that the screening criteria have been fulfilled, or the Conference of the Parties has decided that the proposal should proceed, the Committee shall further review the proposal, taking into account any relevant additional information received, and shall prepare a draft risk profile in accordance with Annex E. It shall, through the Secretariat, make that draft available to all Parties and observers, collect technical comments from them and, taking those comments into account, complete the risk profile.

7. If, on the basis of the risk profile conducted in accordance with Annex E, the Committee decides:

(a) That the chemical is likely as a result of its long-range environmental transport to lead to significant adverse human health and/or environmental effects such that global action is warranted, the proposal shall proceed. Lack of full scientific certainty shall not prevent the proposal from proceeding. The Committee shall, through the Secretariat, invite information from all Parties and observers relating to the considerations specified in Annex F. It shall then prepare a risk management evaluation that includes an analysis of possible control measures for the chemical in accordance with that Annex; or

(b) That the proposal should not proceed, it shall, through the Secretariat, make the risk profile available to all Parties and observers and set the proposal aside.

8. For any proposal set aside pursuant to paragraph 7 (b), a Party may request the Conference of the Parties to consider instructing the Committee to invite additional information from the proposing Party and other Parties during a period not to exceed one year. After that period and on the basis of any information received, the Committee shall reconsider the proposal pursuant to paragraph 6 with a priority to be decided by the Conference of the Parties. If, following this procedure, the Committee again sets the proposal aside, the Party may challenge the decision of the Committee and the Conference of the Parties shall consider the matter at its next session. The Conference of the Parties may decide, based on the risk profile prepared in accordance with Annex E and taking into account the evaluation of the Committee and any additional information provided by any Party or observer, that the proposal should proceed. If the Conference of the Parties decides that the proposal shall proceed, the Committee shall then prepare the risk management evaluation.

9. The Committee shall, based on the risk profile referred to in paragraph 6 and the risk management evaluation referred to in paragraph 7 (a) or paragraph 8, recommend whether the chemical should be considered by the Conference of the Parties for listing in Annexes A, B and/or C. The Conference of the Parties, taking due account of the recommendations of the Committee, including any scientific uncertainty, shall decide, in a precautionary manner, whether to list the chemical, and specify its related control measures, in Annexes A, B and/or C.

Article 9
Information exchange

1. Each Party shall facilitate or undertake the exchange of information relevant to:

(a) The reduction or elimination of the production, use and release of persistent organic pollutants; and

(b) Alternatives to persistent organic pollutants, including information relating to their risks as well as to their economic and social costs.

2. The Parties shall exchange the information referred to in paragraph 1 directly or through the Secretariat.

3. Each Party shall designate a national focal point for the exchange of such information.

4. The Secretariat shall serve as a clearing-house mechanism for information on persistent organic pollutants, including information provided by Parties, intergovernmental organizations and non-governmental organizations.

5. For the purposes of this Convention, information on health and safety of humans and the environment shall not be regarded as confidential. Parties that exchange other information pursuant to this Convention shall protect any confidential information as mutually agreed.

Article 10
Public information, awareness and education

Each Party shall, within its capabilities, promote and facilitate:

(a) Awareness among its policy and decision makers with regard to persistent organic pollutants;

(b) Provision to the public of all available information on persistent organic pollutants, taking into account paragraph 5 of Article 9;

(c) Development and implementation, especially for women, children and the least educated, of educational and public awareness programmes on persistent organic pollutants, as well as on their health and environmental effects and on their alternatives;

(d) Public participation in addressing persistent organic pollutants and their health and environmental effects and in developing adequate responses, including opportunities for providing input at the national level regarding implementation of this Convention;

(e) Training of workers, scientists, educators and technical and managerial personnel;

(f) Development and exchange of educational and public awareness materials at the national and international levels; and

(g) Development and implementation of education and training programmes at the national and international levels.

2. Each Party shall, within its capabilities, ensure that the public has access to the public information referred to in paragraph 1 and that the information is kept up-to-date.

3. Each Party shall, within its capabilities, encourage industry and professional users to promote and facilitate the provision of the information referred to in paragraph 1 at the national level and, as appropriate, subregional, regional and global levels.

4. In providing information on persistent organic pollutants and their alternatives, Parties may use safety data sheets, reports, mass media and other means of communication, and may establish information centres at national and regional levels.

5. Each Party shall give sympathetic consideration to developing mechanisms, such as pollutant release and transfer registers, for the collection and dissemination of information on estimates of the annual quantities of the chemicals listed in Annex A, B or C that are released or disposed of.

Article 11
Research, development and monitoring

1. The Parties shall, within their capabilities, at the national and international levels, encourage and/or undertake appropriate research, development, monitoring and cooperation pertaining to persistent organic pollutants and, where relevant, to their alternatives and to candidate persistent organic pollutants, including on their:

(a) Sources and releases into the environment;

(b) Presence, levels and trends in humans and the environment;

(c) Environmental transport, fate and transformation;

(d) Effects on human health and the environment;

(e) Socio-economic and cultural impacts;

(f) Release reduction and/or elimination; and

(g) Harmonized methodologies for making inventories of generating sources and analytical techniques for the measurement of releases.

2. In undertaking action under paragraph 1, the Parties shall, within their capabilities:

(a) Support and further develop, as appropriate, international programmes, networks and organizations aimed at defining, conducting, assessing and financing research, data collection and monitoring, taking into account the need to minimize duplication of effort;

(b) Support national and international efforts to strengthen national scientific and technical research capabilities, particularly in developing countries and countries with economies in transition, and to promote access to, and the exchange of, data and analyses;

(c) Take into account the concerns and needs, particularly in the field of financial and technical resources, of developing countries and countries with economies in transition and cooperate in improving their capability to participate in the efforts referred to in subparagraphs (a) and (b);

(d) Undertake research work geared towards alleviating the effects of persistent organic pollutants on reproductive health;

(e) Make the results of their research, development and monitoring activities referred to in this paragraph accessible to the public on a timely and regular basis; and

(f) Encourage and/or undertake cooperation with regard to storage and maintenance of information generated from research, development and monitoring.

Article 12
Technical assistance

1. The Parties recognize that rendering of timely and appropriate technical assistance in response to requests from developing country Parties and Parties with economies in transition is essential to the successful implementation of this Convention.

2. The Parties shall cooperate to provide timely and appropriate technical assistance to developing country Parties and Parties with economies in transition, to assist them, taking into account their particular needs, to develop and strengthen their capacity to implement their obligations under this Convention.

3. In this regard, technical assistance to be provided by developed country Parties, and other Parties in accordance with their capabilities, shall include, as appropriate and as mutually agreed, technical assistance for capacity-building relating to implementation of the obligations under this Convention. Further guidance in this regard shall be provided by the Conference of the Parties.

4. The Parties shall establish, as appropriate, arrangements for the purpose of providing technical assistance and promoting the transfer of technology to developing country Parties and Parties with economies in transition relating to the implementation of this Convention. These arrangements shall include regional and subregional centres for capacity-building and transfer of technology to assist developing country Parties and Parties with economies in transition to fulfil their obligations under this Convention. Further guidance in this regard shall be provided by the Conference of the Parties.

5. The Parties shall, in the context of this Article, take full account of the specific needs and special situation of least developed countries and small island developing states in their actions with regard to technical assistance.

Article 13
Financial resources and mechanisms

1. Each Party undertakes to provide, within its capabilities, financial support and incentives in respect of those national activities that are intended to achieve the objective of this Convention in accordance with its national plans, priorities and programmes.

2. The developed country Parties shall provide new and additional financial resources to enable developing country Parties and Parties with economies in transition to meet the agreed full incremental costs of implementing measures which fulfill their obligations under this Convention as agreed between a recipient Party and an entity participating in the mechanism described in paragraph 6. Other Parties may also on a voluntary basis and in accordance with their capabilities provide such financial resources.

Contributions from other sources should also be encouraged. The implementation of these commitments shall take into account the need for adequacy, predictability, the timely flow of funds and the importance of burden sharing among the contributing Parties.

3. Developed country Parties, and other Parties in accordance with their capabilities and in accordance with their national plans, priorities and programmes, may also provide and developing country Parties and Parties with economies in transition avail themselves of financial resources to assist in their implementation of this Convention through other bilateral, regional and multilateral sources or channels.

4. The extent to which the developing country Parties will effectively implement their commitments under this Convention will depend on the effective implementation by developed country Parties of their commitments under this Convention relating to financial resources, technical assistance and technology transfer. The fact that sustainable economic and social development and eradication of poverty are the first and overriding priorities of the developing country Parties will be taken fully into account, giving due consideration to the need for the protection of human health and the environment.

5. The Parties shall take full account of the specific needs and special situation of the least developed countries and the small island developing states in their actions with regard to funding.

6. A mechanism for the provision of adequate and sustainable financial resources to developing country Parties and Parties with economies in transition on a grant or concessional basis to assist in their implementation of the Convention is hereby defined. The mechanism shall function under the authority, as appropriate, and guidance of, and be accountable to the Conference of the Parties for the purposes of this Convention. Its operation shall be entrusted to one or more entities, including existing international entities, as may be decided upon by the Conference of the Parties. The mechanism may also include other entities

Appendix

providing multilateral, regional and bilateral financial and technical assistance. Contributions to the mechanism shall be additional to other financial transfers to developing country Parties and Parties with economies in transition as reflected in, and in accordance with, paragraph 2.

7. Pursuant to the objectives of this Convention and paragraph 6, the Conference of the Parties shall at its first meeting adopt appropriate guidance to be provided to the mechanism and shall agree with the entity or entities participating in the financial mechanism upon arrangements to give effect thereto. The guidance shall address, *inter alia*:

(a) The determination of the policy, strategy and programme priorities, as well as clear and detailed criteria and guidelines regarding eligibility for access to and utilization of financial resources including monitoring and evaluation on a regular basis of such utilization;

(b) The provision by the entity or entities of regular reports to the Conference of the Parties on adequacy and sustainability of funding for activities relevant to the implementation of this Convention;

(c) The promotion of multiple-source funding approaches, mechanisms and arrangements;

(d) The modalities for the determination in a predictable and identifiable manner of the amount of funding necessary and available for the implementation of this Convention, keeping in mind that the phasing out of persistent organic pollutants might require sustained funding, and the conditions under which that amount shall be periodically reviewed; and

(e) The modalities for the provision to interested Parties of assistance with needs assessment, information on available sources of funds and on funding patterns in order to facilitate coordination among them.

8. The Conference of the Parties shall review, not later than its second meeting and thereafter on a regular basis, the effectiveness of the mechanism established under this Article, its ability to address the changing needs of the developing country Parties and Parties with economies in transition, the criteria and guidance referred to in paragraph 7, the level of funding as well as the effectiveness of the performance of the institutional entities entrusted to operate the financial mechanism. It shall, based on such review, take appropriate action, if necessary, to improve the effectiveness of the mechanism, including by means of recommendations and guidance on measures to ensure adequate and sustainable funding to meet the needs of the Parties.

Article 14
Interim financial arrangements

The institutional structure of the Global Environment Facility, operated in accordance with the Instrument for the Establishment of the Restructured Global Environment Facility, shall, on an interim basis, be the principal entity entrusted with the operations of the financial mechanism referred to in Article 13, for the period between the date of entry into force of this Convention and the first meeting of the Conference of the Parties, or until such time as the Conference of the Parties decides which institutional structure will be designated in accordance with Article 13. The institutional structure of the Global Environment Facility should fulfill this function through operational measures related specifically to persistent organic pollutants taking into account that new arrangements for this area may be needed.

Article 15
Reporting

Each Party shall report to the Conference of the Parties on the measures it has taken to implement the provisions of this Convention and on the effectiveness of such measures in meeting the objectives of the Convention.

Each Party shall provide to the Secretariat:

(a) Statistical data on its total quantities of production, import and export of each of the chemicals listed in Annex A and Annex B or a reasonable estimate of such data; and

(b) To the extent practicable, a list of the States from which it has imported each such substance and the States to which it has exported each such substance.

Such reporting shall be at periodic intervals and in a format to be decided by the Conference of the Parties at its first meeting.

Article 16
Effectiveness evaluation

1. Commencing four years after the date of entry into force of this Convention, and periodically thereafter at intervals to be decided by the Conference of the Parties, the Conference shall evaluate the effectiveness of this Convention.

2. In order to facilitate such evaluation, the Conference of the Parties shall, at its first meeting, initiate the establishment of arrangements to provide itself with comparable monitoring data on the presence of the chemicals listed in Annexes A, B and C as well as their regional and global environmental transport. These arrangements:

(a) Should be implemented by the Parties on a regional basis when appropriate, in accordance with their technical and financial capabilities, using existing monitoring programmes and mechanisms to the extent possible and promoting harmonization of approaches;

(b) May be supplemented where necessary, taking into account the differences between regions and their capabilities to implement monitoring activities; and

(c) Shall include reports to the Conference of the Parties on the results of the monitoring activities on a regional and global basis at intervals to be specified by the Conference of the Parties.

3. The evaluation described in paragraph 1 shall be conducted on the basis of available scientific, environmental, technical and economic information, including:

(a) Reports and other monitoring information provided pursuant to paragraph 2;

(b) National reports submitted pursuant to Article 15; and

(c) Non-compliance information provided pursuant to the procedures established under Article 17.

Article 17
Non-compliance

The Conference of the Parties shall, as soon as practicable, develop and approve procedures and institutional mechanisms for determining non-compliance with the provisions of this Convention and for the treatment of Parties found to be in non-compliance.

Article 18
Settlement of disputes

1. Parties shall settle any dispute between them concerning the interpretation or application of this Convention through negotiation or other peaceful means of their own choice.

2. When ratifying, accepting, approving or acceding to the Convention, or at any time thereafter, a Party that is not a regional economic integration organization may declare in a written instrument submitted to the depositary that, with respect to any dispute concerning the interpretation or application of the Convention, it recognizes one or both of the following means of dispute settlement as compulsory in relation to any Party accepting the same obligation:

(a) Arbitration in accordance with procedures to be adopted by the Conference of the Parties in an annex as soon as practicable;

(b) Submission of the dispute to the International Court of Justice.

3. A Party that is a regional economic integration organization may make a declaration with like effect in relation to arbitration in accordance with the procedure referred to in paragraph 2 (a).

4. A declaration made pursuant to paragraph 2 or paragraph 3 shall remain in force until it expires in accordance with its terms or until three months after written notice of its revocation has been deposited with the depositary.

5. The expiry of a declaration, a notice of revocation or a new declaration shall not in any way affect proceedings pending before an arbitral tribunal or the International Court of Justice unless the parties to the dispute otherwise agree.

6. If the parties to a dispute have not accepted the same or any procedure pursuant to paragraph 2, and if they have not been able to settle their dispute within twelve months following notification by one party to another that a dispute exists between them, the dispute shall be submitted to a conciliation commission at the request of any party to the dispute. The conciliation commission shall render a report with recommendations. Additional procedures relating to the conciliation commission shall be included in an annex to be adopted by the Conference of the Parties no later than at its second meeting.

Article 19
Conference of the Parties

1. A Conference of the Parties is hereby established.

2. The first meeting of the Conference of the Parties shall be convened by the Executive Director of the United Nations Environment Programme no later than one year after the entry into force of this Convention. Thereafter, ordinary meetings of the Conference of the Parties shall be held at regular intervals to be decided by the Conference.

3. Extraordinary meetings of the Conference of the Parties shall be held at such other times as may be deemed necessary by the Conference, or at the written request of any Party provided that it is supported by at least one third of the Parties.

4. The Conference of the Parties shall by consensus agree upon and adopt at its first meeting rules of procedure and financial rules for itself and any subsidiary bodies, as well as financial provisions governing the functioning of the Secretariat.

5. The Conference of the Parties shall keep under continuous review and evaluation the implementation of this Convention. It shall perform the functions assigned to it by the Convention and, to this end, shall:

   (a) Establish, further to the requirements of paragraph 6, such subsidiary bodies as it considers necessary for the implementation of the Convention;

   (b) Cooperate, where appropriate, with competent international organizations and intergovernmental and non-governmental bodies; and

   (c) Regularly review all information made available to the Parties pursuant to Article 15, including consideration of the effectiveness of paragraph 2 (b) (iii) of Article 3;

   (d) Consider and undertake any additional action that may be required for the achievement of the objectives of the Convention.

6. The Conference of the Parties shall, at its first meeting, establish a subsidiary body to be called the Persistent Organic Pollutants Review Committee for the

purposes of performing the functions assigned to that Committee by this Convention. In this regard:

(a) The members of the Persistent Organic Pollutants Review Committee shall be appointed by the Conference of the Parties. Membership of the Committee shall consist of government-designated experts in chemical assessment or management. The members of the Committee shall be appointed on the basis of equitable geographical distribution;

(b) The Conference of the Parties shall decide on the terms of reference, organization and operation of the Committee; and

(c) The Committee shall make every effort to adopt its recommendations by consensus. If all efforts at consensus have been exhausted, and no consensus reached, such recommendation shall as a last resort be adopted by a two-thirds majority vote of the members present and voting.

7. The Conference of the Parties shall, at its third meeting, evaluate the continued need for the procedure contained in paragraph 2 (b) of Article 3, including consideration of its effectiveness.

8. The United Nations, its specialized agencies and the International Atomic Energy Agency, as well as any State not Party to this Convention, may be represented at meetings of the Conference of the Parties as observers. Any body or agency, whether national or international, governmental or non-governmental, qualified in matters covered by the Convention, and which has informed the Secretariat of its wish to be represented at a meeting of the Conference of the Parties as an observer may be admitted unless at least one third of the Parties present object. The admission and participation of observers shall be subject to the rules of procedure adopted by the Conference of the Parties.

Article 20
Secretariat

1. A Secretariat is hereby established.

2. The functions of the Secretariat shall be:

(a) To make arrangements for meetings of the Conference of the Parties and its subsidiary bodies and to provide them with services as required;

(b) To facilitate assistance to the Parties, particularly developing country Parties and Parties with economies in transition, on request, in the implementation of this Convention;

(c) To ensure the necessary coordination with the secretariats of other relevant international bodies;

(d) To prepare and make available to the Parties periodic reports based on information received pursuant to Article 15 and other available information;

(e) To enter, under the overall guidance of the Conference of the Parties, into such administrative and contractual arrangements as may be required for the effective discharge of its functions; and

(f) To perform the other secretariat functions specified in this Convention and such other functions as may be determined by the Conference of the Parties.

3. The secretariat functions for this Convention shall be performed by the Executive Director of the United Nations Environment Programme, unless the Conference of the Parties decides, by a three-fourths majority of the Parties present and voting, to entrust the secretariat functions to one or more other international organizations.

Article 21
Amendments to the Convention

1. Amendments to this Convention may be proposed by any Party.

2. Amendments to this Convention shall be adopted at a meeting of the Conference of the Parties. The text of any proposed amendment shall be communicated to the Parties by the Secretariat at least six months before the meeting at which it is proposed for adoption. The Secretariat shall also communicate proposed amendments to the signatories to this Convention and, for information, to the depositary.

3. The Parties shall make every effort to reach agreement on any proposed amendment to this Convention by consensus. If all efforts at consensus have been exhausted, and no agreement reached, the amendment shall as a last resort be adopted by a three-fourths majority vote of the Parties present and voting.

4. The amendment shall be communicated by the depositary to all Parties for ratification, acceptance or approval.

5. Ratification, acceptance or approval of an amendment shall be notified to the depositary in writing. An amendment adopted in accordance with paragraph 3 shall enter into force for the Parties having accepted it on the ninetieth day after the date of deposit of instruments of ratification, acceptance or approval by at least three-fourths of the Parties. Thereafter, the amendment shall enter into force for any other Party on the ninetieth day after the date on which that Party deposits its instrument of ratification, acceptance or approval of the amendment.

Article 22
Adoption and amendment of annexes

1. Annexes to this Convention shall form an integral part thereof and, unless expressly provided otherwise, a reference to this Convention constitutes at the same time a reference to any annexes thereto.

2. Any additional annexes shall be restricted to procedural, scientific, technical or administrative matters.

3. The following procedure shall apply to the proposal, adoption and entry into force of additional annexes to this Convention:

(a) Additional annexes shall be proposed and adopted according to the procedure laid down in paragraphs 1, 2 and 3 of Article 21;

(b) Any Party that is unable to accept an additional annex shall so notify the depositary, in writing, within one year from the date of communication by the depositary of the adoption of the additional annex. The depositary shall without delay notify all Parties of any such notification received. A Party may at any time withdraw a previous notification of non-acceptance in respect of any additional annex, and the annex shall thereupon enter into force for that Party subject to subparagraph (c); and

(c) On the expiry of one year from the date of the communication by the depositary of the adoption of an additional annex, the annex shall enter into force for all Parties that have not submitted a notification in accordance with the provisions of subparagraph (b).

4. The proposal, adoption and entry into force of amendments to Annex A, B or C shall be subject to the same procedures as for the proposal, adoption and entry into force of additional annexes to this Convention, except that an amendment to Annex A, B or C shall not enter into force with respect to any Party that has made a declaration with respect to amendment to those Annexes in accordance with paragraph 4 of Article 25, in which case any such amendment shall enter into force for such a Party on the ninetieth day after the date of deposit with the depositary of its instrument of ratification, acceptance, approval or accession with respect to such amendment.

5. The following procedure shall apply to the proposal, adoption and entry into force of an amendment to Annex D, E or F:

(a) Amendments shall be proposed according to the procedure in paragraphs 1 and 2 of Article 21;

(b) The Parties shall take decisions on an amendment to Annex D, E or F by consensus; and

(c) A decision to amend Annex D, E or F shall forthwith be communicated to the Parties by the depositary. The amendment shall enter into force for all Parties on a date to be specified in the decision.

6. If an additional annex or an amendment to an annex is related to an amendment to this Convention, the additional annex or amendment shall not enter into force until such time as the amendment to the Convention enters into force.

Article 23
Right to vote

1. Each Party to this Convention shall have one vote, except as provided for in paragraph 2.

2. A regional economic integration organization, on matters within its competence, shall exercise its right to vote with a number of votes equal to the number of its member States that are Parties to this Convention. Such an organization shall not exercise its right to vote if any of its member States exercises its right to vote, and vice versa.

Article 24
Signature

This Convention shall be open for signature at Stockholm by all States and regional economic integration organizations on 23 May 2001, and at the United Nations Headquarters in New York from 24 May 2001 to 22 May 2002.

Article 25
Ratification, acceptance, approval or accession

1. This Convention shall be subject to ratification, acceptance or approval by States and by regional economic integration organizations. It shall be open for accession by States and by regional economic integration organizations from the day after the date on which the Convention is closed for signature. Instruments of ratification, acceptance, approval or accession shall be deposited with the depositary.

2. Any regional economic integration organization that becomes a Party to this Convention without any of its member States being a Party shall be bound by all the obligations under the Convention. In the case of such organizations, one or more of whose member States is a Party to this Convention, the organization and its member States shall decide on their respective responsibilities for the performance of their obligations under the Convention. In such cases, the organization and the member States shall not be entitled to exercise rights under the Convention concurrently.

3. In its instrument of ratification, acceptance, approval or accession, a regional economic integration organization shall declare the extent of its competence in respect of the matters governed by this Convention. Any such organization shall also inform the depositary, who shall in turn inform the Parties, of any relevant modification in the extent of its competence.

4. In its instrument of ratification, acceptance, approval or accession, any Party may declare that, with respect to it, any amendment to Annex A, B or C shall enter into force only upon the deposit of its instrument of ratification, acceptance, approval or accession with respect thereto.

Article 26
Entry into force

1. This Convention shall enter into force on the ninetieth day after the date of deposit of the fiftieth instrument of ratification, acceptance, approval or accession.

2. For each State or regional economic integration organization that ratifies, accepts or approves this Convention or accedes thereto after the deposit of the fiftieth instrument of ratification, acceptance, approval or accession, the Convention shall enter into force on the ninetieth day after the date of deposit by such State or regional economic integration organization of its instrument of ratification, acceptance, approval or accession.

3. For the purpose of paragraphs 1 and 2, any instrument deposited by a regional economic integration organization shall not be counted as additional to those deposited by member States of that organization.

Article 27
Reservations

No reservations may be made to this Convention.

Article 28
Withdrawal

1. At any time after three years from the date on which this Convention has entered into force for a Party, that Party may withdraw from the Convention by giving written notification to the depositary.

2. Any such withdrawal shall take effect upon the expiry of one year from the date of receipt by the depositary of the notification of withdrawal, or on such later date as may be specified in the notification of withdrawal.

Article 29
Depositary

The Secretary-General of the United Nations shall be the depositary of this Convention.

Article 30
Authentic texts

The original of this Convention, of which the Arabic, Chinese, English, French, Russian and Spanish texts are equally authentic, shall be deposited with the Secretary-General of the United Nations.

IN WITNESS WHEREOF the undersigned, being duly authorized to that effect, have signed this Convention.

Done at Stockholm on this twenty-second day of May, two thousand and one.

Annex A
ELIMINATION
Part I

Chemical	Activity	Specific exemption
Aldrin* CAS No: 309-00-2	Production	None
	Use	Local ectoparasiticide Insecticide
Chlordane* CAS No: 57-74-9	Production	As allowed for the Parties listed in the Register
	Use	Local ectoparasiticide Insecticide Termiticide Termiticide in buildings and dams Termiticide in roads Additive in plywood adhesives
Dieldrin* CAS No: 60-57-1	Production	None
	Use	In agricultural operations
Endrin* CAS No: 72-20-8	Production	None
	Use	None
Heptachlor* CAS No: 76-44-8	Production	None
	Use	Termiticide Termiticide in structures of houses Termiticide (subterranean) Wood treatment In use in underground cable boxes
Hexachlorobenzene CAS No: 118-74-1	Production	As allowed for the Parties listed in the Register
	Use	Intermediate Solvent in pesticide Closed system site limited intermediate
Mirex* CAS No: 2385-85-5	Production	As allowed for the Parties listed in the Register
	Use	Termiticide
Toxaphene* CAS No: 8001-35-2	Production	None
	Use	None
Polychlorinated Biphenyls (PCB)*	Production	None
	Use	Articles in use in accordance with the provisions of Part II of this Annex

Notes:

(i) Except as otherwise specified in this Convention, quantities of a chemical occurring as unintentional trace contaminants in products and articles shall not be considered to be listed in this Annex;

(ii) This note shall not be considered as a production and use specific exemption for purposes of paragraph 2 of Article 3. Quantities of a chemical occurring as constituents of articles manufactured or already in use before or on the date of entry into force of the relevant obligation with respect to that chemical, shall not be considered as listed in this Annex, provided that a Party has notified the Secretariat that a particular type of article remains in use within that Party. The Secretariat shall make such notifications publicly available;

(iii) This note, which does not apply to a chemical that has an asterisk following its name in the Chemical column in Part I of this Annex, shall not be considered as a production and use specific exemption for purposes of paragraph 2 of Article 3. Given that no significant quantities of the chemical are expected to reach humans and the environment during the production and use of a closed-system site-limited intermediate, a Party, upon notification to the Secretariat, may allow the production and use of quantities of a chemical listed in this Annex as a closed-system site-limited intermediate that is chemically transformed in the manufacture of other chemicals that, taking into consideration the criteria in paragraph 1 of Annex D, do not exhibit the characteristics of persistent organic pollutants. This notification shall include information on total production and use of such chemical or a reasonable estimate of such information and information regarding the nature of the closed-system site-limited process including the amount of any non-transformed and unintentional trace contamination of the persistent organic pollutant-starting material in the final product. This procedure applies except as otherwise specified in this Annex. The Secretariat shall make such notifications available to the Conference of the Parties and to the public. Such production or use shall not be considered a production or use specific exemption. Such production and use shall cease after a ten-year period, unless the Party concerned submits a new notification to the Secretariat, in which case the period will be extended for an additional ten years unless the Conference of the Parties, after a review of the production and use decides otherwise. The notification procedure can be repeated;

(iv) All the specific exemptions in this Annex may be exercised by Parties that have registered exemptions in respect of them in accordance with Article 4 with the exception of the use of polychlorinated biphenyls in articles in use in accordance with the provisions of Part II of this Annex, which may be exercised by all Parties.

Part II
Polychlorinated biphenyls

Each Party shall:

(a) With regard to the elimination of the use of polychlorinated biphenyls in equipment (e.g. transformers, capacitors or other receptacles containing liquid stocks) by 2025, subject to review by the Conference of the Parties, take action in accordance with the following priorities:

(i) Make determined efforts to identify, label and remove from use equipment containing greater than 10 per cent polychlorinated biphenyls and volumes greater than 5 litres;

(ii) Make determined efforts to identify, label and remove from use equipment containing greater than 0.05 per cent polychlorinated biphenyls and volumes greater than 5 litres;

(iii) Endeavour to identify and remove from use equipment containing greater than 0.005 percent polychlorinated biphenyls and volumes greater than 0.05 litres;

(b) Consistent with the priorities in subparagraph (a), promote the following measures to reduce exposures and risk to control the use of polychlorinated biphenyls:

(i) Use only in intact and non-leaking equipment and only in areas where the risk from environmental release can be minimised and quickly remedied;

(ii) Not use in equipment in areas associated with the production or processing of food or feed;

(iii) When used in populated areas, including schools and hospitals, all reasonable measures to protect from electrical failure which could result in a fire, and regular inspection of equipment for leaks;

(c) Notwithstanding paragraph 2 of Article 3, ensure that equipment containing polychlorinated biphenyls, as described in subparagraph (a), shall not be exported or imported except for the purpose of environmentally sound waste management;

(d) Except for maintenance and servicing operations, not allow recovery for the purpose of reuse in other equipment of liquids with polychlorinated biphenyls content above 0.005 per cent;

(e) Make determined efforts designed to lead to environmentally sound waste management of liquids containing polychlorinated biphenyls and equipment contaminated with polychlorinated biphenyls having a polychlorinated biphenyls content above 0.005 per cent, in accordance with paragraph 1 of Article 6, as soon as possible but no later than 2028, subject to review by the Conference of the Parties;

(f) In lieu of note (ii) in Part I of this Annex, endeavour to identify other articles containing more than 0.005 per cent polychlorinated biphenyls (e.g. cable-sheaths, cured caulk and painted objects) and manage them in accordance with paragraph 1 of Article 6;

(g) Provide a report every five years on progress in eliminating polychlorinated biphenyls and submit it to the Conference of the Parties pursuant to Article 15;

(h) The reports described in subparagraph (g) shall, as appropriate, be considered by the Conference of the Parties in its reviews relating to polychlorinated biphenyls. The Conference of the Parties shall review progress towards elimination of polychlorinated biphenyls at five year intervals or other period, as appropriate, taking into account such reports.

Annex B
RESTRICTION
Part I

Chemical	Activity	Acceptable purpose or specific exemption
DDT (1,1,1-trichloro-2,2-bis(4-chlorophenyl)ethane) CAS No: 50-29-3	Production	Acceptable purpose: Disease vector control use in accordance with Part II of this Annex Specific exemption: Intermediate in production of dicofol Intermediate
	Use	Acceptable purpose: Disease vector control in accordance with Part II of this Annex Specific exemption: Production of dicofol Intermediate

Notes:

(i) Except as otherwise specified in this Convention, quantities of a chemical occurring as unintentional trace contaminants in products and articles shall not be considered to be listed in this Annex;

(ii) This note shall not be considered as a production and use acceptable purpose or specific exemption for purposes of paragraph 2 of Article 3. Quantities of a chemical occurring as constituents of articles manufactured or already in use before or on the date of entry into force of the relevant obligation with respect to that chemical, shall not be considered as listed in this Annex, provided that a Party has notified the Secretariat that a particular type of article remains in use within that Party. The Secretariat shall make such notifications publicly available;

(iii) This note shall not be considered as a production and use specific exemption for purposes of paragraph 2 of Article 3. Given that no significant quantities of the chemical are expected to reach humans and the environment during the production and use of a closed-system site-limited intermediate, a Party, upon notification to the Secretariat, may allow the production and use of quantities of a chemical listed in this Annex as a closed-system site-limited intermediate that is chemically transformed in the manufacture of other chemicals that, taking into consideration the criteria in paragraph 1 of Annex D, do not exhibit the characteristics of persistent organic pollutants. This notification shall include information on total production and use of such chemical or a reasonable estimate of such information and information regarding the nature of the closed-system site-limited process including the amount of any non-transformed and unintentional trace contamination of the persistent organic pollutant-starting material in the final product. This procedure applies except as otherwise specified in this Annex. The Secretariat shall make such notifications available to the

Conference of the Parties and to the public. Such production or use shall not be considered a production or use specific exemption. Such production and use shall cease after a ten-year period, unless the Party concerned submits a new notification to the Secretariat, in which case the period will be extended for an additional ten years unless the Conference of the Parties, after a review of the production and use decides otherwise. The notification procedure can be repeated;

(iv) All the specific exemptions in this Annex may be exercised by Parties that have registered in respect of them in accordance with Article 4.

Part II
DDT (1,1,1-trichloro-2,2-bis(4-chlorophenyl)ethane)

1. The production and use of DDT shall be eliminated except for Parties that have notified the Secretariat of their intention to produce and/or use it. A DDT Register is hereby established and shall be available to the public. The Secretariat shall maintain the DDT Register.

2. Each Party that produces and/or uses DDT shall restrict such production and/or use for disease vector control in accordance with the World Health Organization recommendations and guidelines on the use of DDT and when locally safe, effective and affordable alternatives are not available to the Party in question.

3. In the event that a Party not listed in the DDT Register determines that it requires DDT for disease vector control, it shall notify the Secretariat as soon as possible in order to have its name added forthwith to the DDT Register. It shall at the same time notify the World Health Organization.

4. Every three years, each Party that uses DDT shall provide to the Secretariat and the World Health Organization information on the amount used, the conditions of such use and its relevance to that Party's disease management strategy, in a format to be decided by the Conference of the Parties in consultation with the World Health Organization.

5. With the goal of reducing and ultimately eliminating the use of DDT, the Conference of the Parties shall encourage:

(a) Each Party using DDT to develop and implement an action plan as part of the implementation plan specified in Article 7. That action plan shall include:

(i) Development of regulatory and other mechanisms to ensure that DDT use is restricted to disease vector control;

(ii) Implementation of suitable alternative products, methods and strategies, including resistance management strategies to ensure the continuing effectiveness of these alternatives;

(iii) Measures to strengthen health care and to reduce the incidence of the disease.

(b) The Parties, within their capabilities, to promote research and development of safe alternative chemical and non-chemical products, methods and strategies for Parties using DDT, relevant to the conditions of those countries and with the goal of decreasing the human and economic burden of disease. Factors to be promoted when considering alternatives or combinations of alternatives shall include the human health risks and environmental implications of such alternatives. Viable alternatives to DDT shall pose less risk to human health and the environment, be suitable for disease control based on conditions in the Parties in question and be supported with monitoring data.

6. Commencing at its first meeting, and at least every three years thereafter, the Conference of the Parties shall, in consultation with the World Health Organization, evaluate the continued need for DDT for disease vector control on the basis of available scientific, technical, environmental and economic information, including:

(a) The production and use of DDT and the conditions set out in paragraph 2;

(b) The availability, suitability and implementation of the alternatives to DDT; and

(c) Progress in strengthening the capacity of countries to transfer safely to reliance on such alternatives.

7. A Party may, at any time, withdraw its name from the DDT Registry upon written notification to the Secretariat. The withdrawal shall take effect on the date specified in the notification.

Annex C
UNINTENTIONAL PRODUCTION

Part I: Persistent organic pollutants subject to the requirements of Article 5

This Annex applies to the following persistent organic pollutants when formed and released unintentionally from anthropogenic sources:

Chemical
Polychlorinated dibenzo-p-dioxins and dibenzofurans (PCDD/PCDF)
Hexachlorobenzene (HCB) (CAS No: 118-74-1)
Polychlorinated biphenyls (PCB)

Part II: Source categories

Polychlorinated dibenzo-p-dioxins and dibenzofurans, hexachlorobenzene and polychlorinated biphenyls are unintentionally formed and released from thermal processes involving organic matter and chlorine as a result of incomplete combustion or chemical reactions. The following industrial source categories have the potential for comparatively high formation and release of these chemicals to the environment:

(a) Waste incinerators, including co-incinerators of municipal, hazardous or medical waste or of sewage sludge;

(b) Cement kilns firing hazardous waste;

(c) Production of pulp using elemental chlorine or chemicals generating elemental chlorine for bleaching;

(d) The following thermal processes in the metallurgical industry:

    (i) Secondary copper production;

    (ii) Sinter plants in the iron and steel industry;

    (iii) Secondary aluminium production;

    (iv) Secondary zinc production.

Part III: Source categories

Polychlorinated dibenzo-p-dioxins and dibenzofurans, hexachlorobenzene and polychlorinated biphenyls may also be unintentionally formed and released from the following source categories, including:

(a) Open burning of waste, including burning of landfill sites;

(b) Thermal processes in the metallurgical industry not mentioned in Part II;

(c) Residential combustion sources;

(d) Fossil fuel-fired utility and industrial boilers;

(e) Firing installations for wood and other biomass fuels;

(f) Specific chemical production processes releasing unintentionally formed persistent organic pollutants, especially production of chlorophenols and chloranil;

(g) Crematoria;

(h) Motor vehicles, particularly those burning leaded gasoline;

(i) Destruction of animal carcasses;

(j) Textile and leather dyeing (with chloranil) and finishing (with alkaline extraction);

(k) Shredder plants for the treatment of end of life vehicles;

(l) Smouldering of copper cables;

(m) Waste oil refineries.

Part IV: Definitions

1. For the purposes of this Annex:

    (a) "Polychlorinated biphenyls" means aromatic compounds formed in such a manner that the hydrogen atoms on the biphenyl molecule (two benzene rings bonded together by a single carbon-carbon bond) may be replaced by up to ten chlorine atoms; and

    (b) "Polychlorinated dibenzo-p-dioxins" and "polychlorinated dibenzofurans" are tricyclic, aromatic compounds formed by two benzene rings connected by two oxygen atoms in polychlorinated dibenzo-p-dioxins and by one oxygen atom and one carbon-carbon bond in polychlorinated dibenzofurans and the hydrogen atoms of which may be replaced by up to eight chlorine atoms.

2. In this Annex, the toxicity of polychlorinated dibenzo-p-dioxins and dibenzofurans is expressed using the concept of toxic equivalency which measures the relative dioxin-like toxic activity of different congeners of polychlorinated dibenzo-p-dioxins and dibenzofurans and coplanar polychlorinated biphenyls in comparison to 2,3,7,8-tetrachlorodibenzo-p-dioxin. The toxic equivalent factor values to be used for the purposes of this Convention shall be consistent with accepted international standards, commencing with the World Health Organization 1998 mammalian toxic equivalent factor values for polychlorinated dibenzo-p-dioxins and dibenzofurans and coplanar polychlorinated biphenyls. Concentrations are expressed in toxic equivalents.

Part V: General guidance on best available techniques and best environmental practices

This Part provides general guidance to Parties on preventing or reducing releases of the chemicals listed in Part I.

A. General prevention measures relating to both best available techniques and best environmental practices

Priority should be given to the consideration of approaches to prevent the formation and release of the chemicals listed in Part I. Useful measures could include:

(a) The use of low-waste technology;

(b) The use of less hazardous substances;

(c) The promotion of the recovery and recycling of waste and of substances generated and used in a process;

(d) Replacement of feed materials which are persistent organic pollutants or where there is a direct link between the materials and releases of persistent organic pollutants from the source;

(e) Good housekeeping and preventive maintenance programmes;

(f) Improvements in waste management with the aim of the cessation of open and other uncontrolled burning of wastes, including the burning of landfill sites. When considering proposals to construct new waste disposal facilities, consideration should be given to alternatives such as activities to minimize the generation of municipal and medical waste, including resource recovery, reuse, recycling, waste separation and promoting products that generate less waste. Under this approach, public health concerns should be carefully considered;

(g) Minimization of these chemicals as contaminants in products;

(h) Avoiding elemental chlorine or chemicals generating elemental chlorine for bleaching.

B. Best available techniques

The concept of best available techniques is not aimed at the prescription of any specific technique or technology, but at taking into account the technical characteristics of the installation concerned, its geographical location and the local environmental conditions. Appropriate control techniques to reduce releases of the chemicals listed in Part I are in general the same. In determining best available techniques, special consideration should be given, generally or in specific cases, to the following factors, bearing in mind the likely costs and benefits of a measure and consideration of precaution and prevention:

(a) General considerations:

(i) The nature, effects and mass of the releases concerned: techniques may vary depending on source size;

(ii) The commissioning dates for new or existing installations;

(iii) The time needed to introduce the best available technique;

(iv) The consumption and nature of raw materials used in the process and its energy efficiency;

(v) The need to prevent or reduce to a minimum the overall impact of the releases to the environment and the risks to it;

(vi) The need to prevent accidents and to minimize their consequences for the environment;

(vii) The need to ensure occupational health and safety at workplaces;

(viii) Comparable processes, facilities or methods of operation which have been tried with success on an industrial scale;

(ix) Technological advances and changes in scientific knowledge and understanding.

(b) General release reduction measures: When considering proposals to construct new facilities or significantly modify existing facilities using processes that release chemicals listed in this Annex, priority consideration should be given to alternative processes, techniques or practices that have similar usefulness but which avoid the formation and release of such chemicals. In cases where such facilities will be constructed or significantly modified, in addition to the prevention measures outlined in section A of Part V the following reduction measures could also be considered in determining best available techniques:

(i) Use of improved methods for flue-gas cleaning such as thermal or catalytic oxidation, dust precipitation, or adsorption;

(ii) Treatment of residuals, wastewater, wastes and sewage sludge by, for example, thermal treatment or rendering them inert or chemical processes that detoxify them;

(iii) Process changes that lead to the reduction or elimination of releases, such as moving to closed systems;

(iv) Modification of process designs to improve combustion and prevent formation of the chemicals listed in this Annex, through the control of parameters such as incineration temperature or residence time.

C. Best environmental practices

The Conference of the Parties may develop guidance with regard to best environmental practices.

Annex D
INFORMATION REQUIREMENTS AND SCREENING CRITERIA

1. A Party submitting a proposal to list a chemical in Annexes A, B and/or C shall identify the chemical in the manner described in subparagraph (a) and provide the information on the chemical, and its transformation products where relevant, relating to the screening criteria set out in subparagraphs (b) to (e):

(a) *Chemical identity*:

(i) Names, including trade name or names, commercial name or names and synonyms, Chemical Abstracts Service (CAS) Registry number, International Union of Pure and Applied Chemistry (IUPAC) name; and

(ii) Structure, including specification of isomers, where applicable, and the structure of the chemical class;

(b) *Persistence*:

(i) Evidence that the half-life of the chemical in water is greater than two months, or that its half-life in soil is greater than six months, or that its half-life in sediment is greater than six months; or

(ii) Evidence that the chemical is otherwise sufficiently persistent to justify its consideration within the scope of this Convention;

(c) *Bio-accumulation*:

(i) Evidence that the bio-concentration factor or bio-accumulation factor in aquatic species for the chemical is greater than 5,000 or, in the absence of such data, that the log Kow is greater than 5;

(ii) Evidence that a chemical presents other reasons for concern, such as high bio-accumulation in other species, high toxicity or ecotoxicity; or

(iii) Monitoring data in biota indicating that the bio-accumulation potential of the chemical is sufficient to justify its consideration within the scope of this Convention;

(d) *Potential for long-range environmental transport*:

(i) Measured levels of the chemical in locations distant from the sources of its release that are of potential concern;

(ii) Monitoring data showing that long-range environmental transport of the chemical, with the potential for transfer to a receiving environment, may have occurred via air, water or migratory species; or

(iii) Environmental fate properties and/or model results that demonstrate that the chemical has a potential for long-range environmental transport through air, water or migratory species, with the potential for transfer to a receiving environment in locations distant from the sources of its release. For a chemical that migrates significantly through the air, its half-life in air should be greater than two days; and

(e) *Adverse effects*:

(i) Evidence of adverse effects to human health or to the environment that justifies consideration of the chemical within the scope of this Convention; or

(ii) Toxicity or ecotoxicity data that indicate the potential for damage to human health or to the environment.

2. The proposing Party shall provide a statement of the reasons for concern including, where possible, a comparison of toxicity or ecotoxicity data with detected or predicted levels of a chemical resulting or anticipated from its long-range environmental transport, and a short statement indicating the need for global control.

3. The proposing Party shall, to the extent possible and taking into account its capabilities, provide additional information to support the review of the proposal referred to in paragraph 6 of Article 8. In developing such a proposal, a Party may draw on technical expertise from any source.

Annex E
INFORMATION REQUIREMENTS FOR THE RISK PROFILE

The purpose of the review is to evaluate whether the chemical is likely, as a result of its long-range environmental transport, to lead to significant adverse human health and/or environmental effects, such that global action is warranted. For this purpose, a risk profile shall be developed that further elaborates on, and evaluates, the information referred to in Annex D and includes, as far as possible, the following types of information:

(a) Sources, including as appropriate:

    (i) Production data, including quantity and location;

    (ii) Uses; and

    (iii) Releases, such as discharges, losses and emissions;

(b) Hazard assessment for the endpoint or endpoints of concern, including a consideration of toxicological interactions involving multiple chemicals;

(c) Environmental fate, including data and information on the chemical and physical properties of a chemical as well as its persistence and how they are linked to its environmental transport, transfer within and between environmental compartments, degradation and transformation to other chemicals. A determination of the bio-concentration factor or bio-accumulation factor, based on measured values, shall be available, except when monitoring data are judged to meet this need;

(d) Monitoring data;

(e) Exposure in local areas and, in particular, as a result of long-range environmental transport, and including information regarding bio-availability;

(f) National and international risk evaluations, assessments or profiles and labelling information and hazard classifications, as available; and

(g) Status of the chemical under international conventions.

Annex F
INFORMATION ON SOCIO-ECONOMIC CONSIDERATIONS

An evaluation should be undertaken regarding possible control measures for chemicals under consideration for inclusion in this Convention, encompassing the full range of options, including management and elimination. For this purpose, relevant information should be provided relating to socio-economic considerations associated with possible control measures to enable a decision to be taken by the Conference of the Parties. Such information should reflect due regard for the differing capabilities and conditions among the Parties and should include consideration of the following indicative list of items:

(a) Efficacy and efficiency of possible control measures in meeting risk reduction goals:

(i) Technical feasibility; and

(ii) Costs, including environmental and health costs;

(b) Alternatives (products and processes):

(i) Technical feasibility;

(ii) Costs, including environmental and health costs;

(iii) Efficacy;

(iv) Risk;

(v) Availability; and

(vi) Accessibility;

(c) Positive and/or negative impacts on society of implementing possible control measures:

(i) Health, including public, environmental and occupational health;

(ii) Agriculture, including aquaculture and forestry;

(iii) Biota (biodiversity);

(iv) Economic aspects;

(v) Movement towards sustainable development; and

(vi) Social costs;

(d) Waste and disposal implications (in particular, obsolete stocks of pesticides and clean-up of contaminated sites):

(i) Technical feasibility; and

(ii) Cost;

(e) Access to information and public education;

(i) Status of control and monitoring capacity; and

(ii) Any national or regional control actions taken, including information on alternatives, and other relevant risk management information.

GLOSSARY OF ACRONYMS

AEPS	Arctic Environmental Protection Strategy
AMAP	Arctic Monitoring and Assessment Program
CAS	Chemical Abstract Service
CEG	Criteria Expert Group
COP	Conference of the Party
CPSC	Consumer Product Safety Commission
DDT	Dichlorodiphenyltrichloroethane
ECOSOC	United Nations Economic and Social Council
ESCAP	United Nations Economic and Social Commission for Asia and the Pacific
EPA	Environmental Protection Agency
FAO	Food and Agriculture Organization
FDA	United States Food and Drug Administration
FDCA	Federal Food, Drug, and Cosmetic Act (1938)
FEPCA	Federal Environmental Pesticides Control Act
FIFRA	Federal Insecticide, Fungicide, and Rodenticide Act (1947)
GEF	Global Environment Facility
GRULCA	Group of Latin American and Caribbean Countries
HCH	Hexachlorocyclohexane (CAS: 608-73-1)
IARC	International Agency for Research on Cancer
IFCS	Intergovernmental Forum on Chemical Safety
IGO	Inter Governmental Organization
ILO	International Labor Organization
INC	Intergovernmental Negotiating Committee
IOMC	Inter-Organization Program for the Sound Management of Chemicals
IPCS	International Program on Chemical Safety
IRPTC	International Register of Potentially Toxic Chemicals
IUCN	International Union for Conservation of Nature and Natural Resources (World Conservation Union)

IUPA	Internal Union of Pure and Applied Chemistry
LRTAP	Convention on the Long-Range Transboundary Air Pollution
LD50	Lethal dose for 50% survival
NGO	Non-Governmental Organization
OAU	Organization of African States
OECD	Organization for Economic Cooperation and Development
ODS	Ozone Depleting Substances
PAN	Pesticide Action Network
PBTs	Persistent, Bioaccumulative, Toxic Substances
PCBs	Polychlorinated biphenyls
POPs	Persistent Organic Pollutants
PCDD	Polychlorodibenzo-p-dioxin
PCDF	Polychlorodibenzefuran
PIC	Prior Informed Consent
Ppb	Parts per billion
TSCA	Toxic Substances Control Act
TCDD	Tetrachlorodibenzo-p-dioxin. (The most synthetic substance known.)
UNDP	United Nations Development Programme
UNCED	United Nations Conference on the Environment and Development
UNECE	United Nations Economic Commission for Europe
UNEP	United Nations Environmental Program
UNIDO	United Nations Industrial Development Organization
UNITAR	United Nations Institute for Training and Research
USDA	United States Department of Agriculture
WHO	World Health Organization
WWF	World Wildlife Fund

GLOSSARY OF TERMS

Acute Toxicity—It is the basis for pesticide classification on product labels and it serves to inform users of the potential hazards associated with the use of a particular pesticide. It means there will be immediate effects to an exposure in a short period of time.

Agent Orange—A half-and-half mixture herbicides 2,4,5-trichlorophenoxyacetic acid and 2,4-dichlorophenoxyacetic acid, used during the Vietnam War.

Algicide—chemical substance to control algae in lakes, canals, swimming pools, water tanks, and other sites.

Antifouling agents—chemical substance used to kill or repel organisms that attach to underwater surfaces, such as boat bottoms.

Antimicrobial—Chemical substance used to kill microorganisms such as bacteria and viruses.

Aplastic Anaemia—a serious and usually lethal blood disorder, in which all cellular elements of bone and marrow are reduced in number due to failure of blood cell precursors to reproduce.

Aromatic—Type of hydrocarbon, typified by benzene. The name comes from its usually strong odor.

Biocide—Chemical substance used to kill microorganisms

Bioaccumulation—The absorption and concentration of toxic chemicals in living organisms. Heavy metals and pesticides, such as DDT, are stored in the fatty tissues of animals and passed along to predators of those animals. The result is higher concentration of the pesticide in fatty tissue, eventually reaching harmful levels in predators at the top of the food chain, such as eagles. Also called biomagnification.

Biodegradable—Capable of being decomposed into natural substances, such as carbon dioxide and water, by biological processes, especially bacterial action.

Biomagnification—Increase in the concentration of a substance through a food chain, for example, from prey to predator.

Carcinogen—A chemical, physical, or biological substance that is capable of causing cancer.

Chemical Abstracts Services—CAS Number, or CAS Registry Number, or CAS RN, or CAS#. It is a unique numeric identifier, which designates only one substance and has no chemical significance, assigned by the Chemical Abstracts Service, a division of the American Chemical Society. This is the largest and most current database of chemical substance information in the world containing more than 32,000,000 substance record and approximately 4,000 new substances are added each day.

Chlorinated Hydrocarbon Insecticides—A notorious group of insecticides containing carbon, hydrogen, and chlorine; they act as nerve poisons.

Chronic Toxicity—It is the amount of a pesticide that will cause injury during repeated exposure over a period of time.

Dioxin—A class of chemical compounds containing carbon, hydrogen, oxygen and chlorine, part of a large group called polycyclic halogenated aromatics. Dioxins are human-made by-products of industrial processes such as incineration, paper milling, pesticide manufacture, and smelting.

Disinfectants and Sanitizers—Chemical substance used to kill or inactivate disease-producing microorganisms on inanimate objects.

Dissipation—The disappearance of a substance and is a combination of at least two processes, degradation and mobility.

Distribution Coefficient (Kd)—A measure of how sorbed a substance is to soil or sediment particles, expressed as the ratio of the sorbed-phase concentration to the solution-phase concentration at equilibrium.

Endocrine Disrupter—A substance that affects the endocrine, that is hormone-producing, function, thus causing adverse health effects on an organism or its offspring.

Endometriosis—is a painful, chronic gynecological disorder in which uterine tissues grow outside the uterus.

Filarial Worm—A group of long, hair like nematodes in which the adults live in the blood tissues of vertebrates, causing diseases such as *Elephantiasis* and *River Blindness*.

Fungicide—Any substance that kills fungi (including blights, mildews, molds, and rusts).

Fumigants—Any substance used to produce gas or vapor intended to destroy pests in building or soil.

Gas Chromotography—A method of analysis which the components of a chemical mixture can be separated, identified, or purified. Also called gas-liquid chromatography (GLC), gas-solid chromatography (GSC), or vapor-phase chromatography (VPC).

Halogen—A non-metallic element, such as fluorine, chlorine, bromine, iodine, or astatine.

Herbicide—Any substance used to kill unwanted plants. Herbicides work in different ways; some sterilize the soil, such as those used to keep railroad rights-of-way clear of all growth, others prevent seeds from sprouting, others kill plants once they have sprouted. Some are plant hormones that disrupt the growth control mechanism of the plant.

Insecticide—Any natural or synthetic compound (pesticide) used to kill insects.

Isomer—Any of two chemical substances composed of the same elements in the same proportions but with different structures and different properties. Dioxins and PCBs have many isomers.

K_{ow}—Octanol-water Partition Coefficient is a way of measuring a substance's propensity to bioconcentrate.

Malaria—Common tropical disease cause by four different species of a parasite, *Plasmodium* that lives in human red blood cells. It enters the blood through the bite of certain mosquitoes (of the genus *Anopheles*) that serve as its alternate host. It is characterized by recurring fever. Also called paludism or swamp fever.

Microorganism—a microscopic or submicroscopic organism, one that is too small to be seen by the naked eye. Bacteria, viruses, protozoan, and some fungi and algae are examples of microorganisms.

Molluscicides—Any substance used to kill snails and slugs.

Mutagenic—Capable of changing the structure of DNA without killing the affected cell.

Nematicides—Any substance used to kill nematodes, which are invertebrate animals of the phylum nemathelminthes and class nematoda; these are unsegmented round worms inhabiting soil, water, plants, or plant parts that feed on plant roots.

Onchocerciasis—Skin or eye disease caused by infestation with nematodes (small, wormlike organisms of the genus *Onchocerca*) that are found in West Africa. One species causes skin swellings around cysts containing the parasite; another species infects the eye and can cause blindness.

Organochlorine compounds—Another term for chlorinated hydrocarbons.

Ovicide—Substance used to kill eggs of insects and mites.

Persistence—The length of time that a chemical substance, a pesticide for instance, remains in the environment, whether it stays where it was poured or moves through air, soil, water, or living organism.

Persistent Pesticide—Chemical compounds used for pest control that do not readily break down once released into the environment. They become more or less permanent features of the ecosystem, working their way up the food chain to reach high concentrations in the tissues of higher predators.

Pest—An animal or plant that is directly or indirectly detrimental to human interests, causing harm or reducing the quality and value of a harvestable crop or other resource. Weeds, termites, and rats are examples of pests.

Pesticide—any substance or mixture of substances intended for preventing, destroying, repelling, or mitigating any pest. Any physical, chemical, or biological agent that will kill an undesirable plant or animal pest.

Pheromone—Biochemicals used to disrupt the mating behavior of insects.

Photolytic—A process of chemical decomposition by the action of radiant energy.

Polycyclic Aromatic Hydrocarbon (PAH)—A group of aromatic hydrocarbons having three or more aromatic nuclei in their structures.

Residue—Material remaining after some process has occurred, such as pesticide residues that stay in the soil after the pests have been killed.

Rodenticide—Substance used to control mice and other rodents.

Schistosomiasis—Human parasitic disease (sometimes called bilharziasis or snail fever) caused by three main species of worms: *Schistosoma haematobium*, *S. japonicum*, and *S. mansoni*. It ranks second behind malaria in terms of socio-economic and public health importance in tropical and subtropical areas. According to the World Health Organization it is endemic in 74 developing countries, infecting more than 20 million people in rural areas.

Solubility—The readiness with which a substance dissolves in another substance.

Stability—The inherent ability of an ecosystem (or any system) to resist change, or to maintain steady-state conditions when confronted by a disturbance.

Strychnine—A chemical compound (CAS no.: 57-24-9, Chemical Formula: $C_{21}H_{22}N_2O_2$) which is extremely toxic and may be fatal if swallowed or inhaled. It causes irritation to skin, eyes, and respiratory tract and affects the nervous system.

Synthetic Chemicals—Compounds produced in the laboratory or in large-scale chemical plants as opposed to being extracted from living organisms.

Temephos—This chemical compound (CAS no.: 3383-96-8, Chemical Formula: $C_{16}H_{20}O_6P_2S_3$) is an organophosphate insecticide used for the control of aquatic larvae of mosquitos, midges, gnats, punkies, and sandflies. Its trade name is Abate®.

Teratogenic—An agent or factor, such as radiation, that is capable of inducing developmental malformation of the fetus.

Toxicant—is a toxic substance that is produced by or is a by-product of human made activities.

Toxicity—a) The potency of a poisonous substance, the degree to which it is harmful to organisms. b) The amount of poison found in a substance or produced by an organism.

Toxicity Class I—These are highly toxic pesticides that require specific safety measures. A taste to a teaspoon can kill a person.

Toxicity Class II—These are moderately toxic substances meaning a warning.

Toxicity Class III and IV—These are slightly toxic or relatively nontoxic substances, requiring basic safety precautions.

Toxicology—The study of the adverse effects of chemicals on living organisms.

Toxicologist—A trained professional in the field of chemistry to examine the nature of those effects (including their cellular, biochemical, and molecular mechanisms of action) and assess the probability of their occurrence.

Toxin—A toxic substance that is produced naturally.

Typhus—An epidemic disease caused by the bacterium *Rickettsia prowazeki*, transmitted to humans by the bite of lice. High fever, weakness, headache, and sometimes a rash characterize it; it is often fatal.

Volatility—The ability of a chemical substance to evaporate into the air.

Volatile Organic Compounds—Hydrocarbon Compounds that have low boiling points, usually less than 100°C, and therefore evaporate readily.

Source:
The Dictionary of Ecology and Environmental Science
Henry W. Art, General Editor (1993).
Henry Holt Reference Book, New York, NY. USA.

INDEX

Acceptance, 79, 90, 116, 118-119, 124, 147-150

Accession, 90, 119, 148-150

Agenda 21, 13, 44, 50, 55-57, 59-60, 71, 77-79, 127
  Chapter 17, 56, 77, 79
  Chapter 19, 56-57, 59-60, 78

Aldrin, 2-3, 14-20, 22, 25, 27-28, 30, 32, 34, 37, 44-45, 48-49, 55, 60, 69-70, 74, 79, 96, 103, 151

Anthropogenic Chemical Compounds
  See Chemicals

Arctic, 3, 28, 30, 36, 39-40, 69, 108, 127

Audubon Societies, 103

Awareness, xi, 11, 13, 43, 45-47, 51, 56, 59, 64, 80, 83-84, 88, 107, 114, 123, 132, 138

Bamako Convention on the Ban of the Import into Africa and the Control of Transboundary Movement of Hazardous Wastes within Africa, 44, 66-67

Basel Convention on the Control of Transboundary Movement of Hazardous Wastes and Their Disposal, xi, 13, 49, 63-66, 112, 122, 127, 135

Biomagnification, 4, 28, 69, 108, 127

By-product, 34-35, 82, 95, 97, 100, 110-111

Canada, xi, 16, 40, 49, 69, 73, 83-84, 87, 97-98, 100, 103, 118-119, 123

Carson, Rachel, 1, 19, 24-25, 43, 121

CBD
  See Convention on Biological Diversity

CEG
  See Criteria Experts Group

Charter of the United Nations, 62, 104-105, 127

Chemical Compounds
  See Chemicals

Chemical Inspectorate, 45, 47

Chemical Substances
  See Chemicals

Chemicals, xi-xii, 1-5, 8-9, 12-23, 25-39, 41, 43-45, 47-52, 54-65, 69-75, 77-91, 94-100, 103, 107, 109-117, 121-123, 125, 127-137, 139, 143, 146, 151-152, 154, 156-164

Chlordane, 2-3, 17-22, 25, 27-28, 30, 32, 34, 37, 39, 44, 48-49, 55, 61, 69-70, 74, 79, 96, 103, 107, 122, 151

Civil Society
  See Non-Government Organizations

Code of Conduct
  See International Code of Conduct on the Distribution and Use of Pesticides

Code of Ethics on the International Trade in Chemicals, 50, 57-58, 79

Commission on Sustainable Development of the United Nations, 58

CONAMA, 48-49

Conference of the Parties, 74, 89-90, 92, 94, 103, 108-109, 111-113, 115-118, 123, 131-133, 135-137, 140-147, 152-153, 155-156, 160, 164

Contamination, xi, 6, 19, 32, 35-36, 39-40, 44, 63, 87, 122, 125, 127, 152, 154

Convention on Biological Diversity, 101, 108, 115-117, 119

Convention on Long-range Transboundary Air Pollution, 14, 67-70

Convention on the Prior Informed Consent Procedures for Certain Hazardous Chemicals and Pesticides in International Trade, 14, 53-55, 63, 71-75, 96, 122, 127

COP
  See Conference of the Parties

Criteria Experts Group, 86, 89-91, 93

Customary Law, 61-63, 108

DDT, xii, 2-3, 9-10, 17-20, 22-25, 27-28, 30, 32, 34, 37, 39, 44-45, 48-49, 55, 60, 69-70, 74, 79, 84, 87, 96, 102, 107, 110, 117, 122-123, 154-156

Dichlorodiphenyltrichloroethane
  See DDT

INDEX

Dieldrin, 2-3, 15-20, 22, 25, 27-28, 30, 32, 34, 37, 39, 44-45, 48-49, 55, 60, 69-70, 74, 79, 96, 103, 151

Dioxins, 3, 5, 67, 69

Dirty Dozen
   See Persistent Organic Pollutants

Earth Summit
   See United Nations Conference on the Environment and Development

ECOSOC
   See United Nations Economic and Social Council

Eliminate POPs, 44, 70, 99, 151

Endrin, 2-3, 15, 17-20, 22, 25, 27-28, 30, 32, 34, 37, 39, 48-49, 61, 69-70, 79, 96, 103, 151

Entry into force, 63, 90, 101-102, 118-119, 131-134, 143, 145, 148, 150, 152, 154

Environmental Code, 46-47

Environmental Law, 13, 32, 47, 57, 61, 87, 104-105, 108-109, 121, 123, 125

Environmental Protection Agency, 11, 16, 21-22, 25, 28, 32, 36, 39, 55

Environmentally Sound Management of Toxic Chemicals Including Prevention of Illegal International Trade in Traffic in Toxic and Dangerous Products
   See Chapter 19 of Agenda 21

EPA
   See Environmental Protection Agency

Esbjerg Declaration, 45

FAO
   See Food and Agriculture Organization

Fertilizers, 1, 77

Financial Resources and Mechanisms, 98, 101, 115, 141

Food and Agriculture Organization, xi, 16, 22, 39, 50, 52-55, 58-59, 63, 71-72, 79, 84-85, 101

Food supply, 1, 8, 25

Furans, 3, 34, 67, 69, 158

GATT
   See General Agreement on Tariffs and Trade

GEF
   See Global Environment Facility

General Agreement on Tariffs and Trade, 90

Global Environment Facility, 84-85, 101, 103, 115-116, 142

Governing Council Decision 19/13C, 83

Great Lakes Binational Toxics Strategy, 49

Greenpeace, 6-8, 84, 87, 90, 104

Hard Law, 50, 61, 63

Hazardous Substances
   See Chemicals

Hazardous Waste
   See Basel Convention on the Control of Transboundary Movement of Hazardous Wastes and Their Disposal

Heptachlor, 2-3, 17-22, 25-28, 30, 32, 34, 37, 39, 48-49, 55, 61, 69-70, 74, 79, 96, 103, 107, 111, 151

Hexachlorobenzene, 2-3, 17-20, 22, 25, 27-28, 30, 32-34, 37, 41, 45, 49, 55, 61, 69, 71, 74, 79, 96, 103, 151, 157

Human Health, xi-xii, 4, 9, 12, 19, 24-25, 37-40, 44, 46, 50, 53-56, 60, 63, 67, 73-74, 78, 82-84, 90-91, 94, 96-97, 99-100, 107-108, 110, 112-113, 116-117, 121-123, 125, 127-128, 130, 134, 137, 139, 141, 156, 162-163

Human Rights, 104, 108

ICJ
   See International Court of Justice

IFCS
   See Intergovernmental Forum on Chemical Safety

ILO
   See International Labor Organization

ILO Chemical Convention
   See International Labor Organization's Chemical Convention

INDEX

INC
See Intergovernmental Negotiating Committee

Indigenous People, 40, 108

Industrial By-products, 3, 91

Information Exchange, 43, 53-54, 57, 59, 73-74, 83, 107, 114, 124, 137

Insecticides, 8-11, 15, 17, 19-21, 23, 25-26, 28-29, 37-38, 151

Insects
See Pests

Intergovernmental Forum on Chemical Safety, 2, 50, 57-58, 60, 68, 79-80, 82-83, 88

Intergovernmental Negotiating Committee, 19, 56, 72, 79, 83-91, 93-95, 97-103, 105, 107, 109, 119

Interim Financial Arrangements, 115-116, 142

International Code of Conduct on the Distribution and Use of Pesticides, 50, 52-55, 58, 71-72, 79

International Court of Justice, 61, 117, 144

International Labor Organization, 59-60, 63, 65, 85

International Labor Organization's Chemical Convention, 63

International POPs Elimination Network, 84, 90

International Program on Chemical Safety, 2, 60, 79

International Register of Potentially Toxic Chemicals, 49-53, 79

International Union for the Conservation of Nature, 84, 104

Inter-Organization Program for the Sound Management of Chemicals, 2, 50, 59-60, 79, 85

Inuit
See Indigenous People

IOMC
See Inter-Organization Program for the Sound Management of Chemicals

IPCS
See International Program on Chemical Safety

IPEN
See International POPs Elimination Network

IRPTC
See International Register of Potentially Toxic Chemicals

IUCN
See International Union for the Conservation of Nature

Law of the Sea, 81

Law of the Treaties
See Vienna Convention on the Law of the Treaties

London Guidelines for the Exchange of Information on Chemicals in International Trade, iv, 50, 54-55, 58, 71, 79, 180

LRTAP
See Convention on Long-range Transboundary Air Pollution

Mirex, 2-3, 17-20, 22, 25, 27-28, 30, 32, 34, 37, 48-49, 61, 69-70, 79, 96, 103, 151

Montreal Protocol on Substances that Deplete the Ozone Layer, 85

Multilateral Treaties, 63

Basel Convention on the Control of Transboundary Movement of Hazardous Wastes and Their Disposal, xi, 13, 49, 63-66, 112, 122, 127, 135

Convention on Biological Diversity, 101, 108, 115-117, 119

Convention on Long-range Transboundary Air Pollution, 14, 67-70

Convention on the Prior Informed Consent Procedures for Certain Hazardous Chemicals and Pesticides in International Trade, 14, 54-55, 71-74, 96, 122, 127

General Agreement on Tariffs and Trade, 90

Montreal Protocol on Substances that Deplete the Ozone Layer, 85

INDEX

Stockholm Convention on Persistent Organic Pollutants, xi-14, 43, 46-49, 52, 55-56, 60, 62-63, 70-71, 77-78, 89-90, 98-103, 105, 107-108, 118-119, 122-125, 127

National Council on the Environment
See CONAMA

NGOs
See Non-Government Organizations

Non-compliance, 117, 123, 144

Non-Government Organizations, xii, 51-52, 72, 84-85, 87, 90, 95, 97-98, 102, 104-105, 108, 114

Objective, 2, 40, 46, 49, 52, 57-60, 66, 70, 86, 99-100, 109, 117, 121, 123-124, 128, 141

OECD
See Organization for Economic Cooperation and Development

Organization for Economic Cooperation and Development, 43, 59, 73, 85

Organochlorinated Compounds, 2, 48

PCBs, 2-3, 12, 19, 30-32, 40-41, 45, 48-49, 55, 64-65, 69-70, 74, 79, 88, 95-96, 103, 112, 122, 151-153, 157-158

Persistent Organic Pollutants, xi-xii, 1-7, 13-14, 19-20, 23, 37-41, 43-45, 47-52, 55, 57, 60-63, 65, 67-70, 73, 77, 79-91, 94-105, 107-109, 111-118, 121-125, 127-128, 130, 134-140, 142-143, 145-146, 152, 154, 157-159

Persistent Organic Pollutants Review Committee, 91, 94, 102, 109, 111, 113, 118, 136, 145-146

Pest Control
See Chemicals

Pesticides, xi-xii, 1-3, 8-11, 14-15, 19-20, 25-26, 28, 32, 34, 36-39, 43-45, 47-48, 50, 52-57, 59, 61, 72, 74-75, 77-79, 82-83, 95-97, 111-112, 115, 117, 121-122, 124, 127-128, 130, 164
See specific pesticide

Pests, xii, 1, 9-10, 17, 23-24, 29, 50, 121

PIC
See Prior Informed Consent

PIC Convention
See Convention on the Prior Informed Consent Procedures for Certain Hazardous Chemicals and Pesticides in International Trade

Polluter Pays Principle, 108-109

Polychlorinated biphenyls
See PCBs

Polychlorinated dibenzodioxins
See Dioxins

Polychlorinated dibenzofurans
See Furans

POPs
See Persistent Organic Pollutants

POPs Convention
See Stockholm Convention on Persistent Organic Pollutants

Potentially Toxic Chemicals
See Persistent Organic Pollutants

Precautionary Approach, 61, 63, 85, 99-100, 108-109, 114, 124, 128, 137

Precautionary Principle, 61, 100, 109, 114, 124

Principle 21
See Customary Law

Prior Informed Consent, 14, 16, 47, 53-55, 58, 63, 71-75, 96, 112, 122, 127, 129

Public Health, 1, 5, 9-10, 16, 18, 24, 50, 53, 62, 79, 84, 90, 96, 125, 127, 159

Public Information, 107, 114, 123, 138

Ratification, 63-64, 73, 90, 103, 111, 119, 122, 147-150

Research, Development and Monitoring, 115, 139-140

Reservations, 62, 119, 150

Rio Declaration, 55-57, 77, 99-100, 108-109, 127-128

INDEX

Rotterdam Convention
 See Convention on the Prior Informed Consent Procedures for Certain Hazardous Chemicals and Pesticides in International Trade

Scientific Uncertainty
 See Precautionary Approach

Secretariat, 55, 60, 62, 70, 72, 84-86, 88-91, 95, 99, 102-103, 105, 111, 113-114, 116, 118, 130-131, 136-138, 143, 145-147, 152, 154-156

Settlement of Disputes, 117, 144

Sierra Club, 84, 103

Signature, xi, 68, 73, 90, 103, 111, 117-118, 122, 149

Silent Spring, 1, 19, 24-25, 43-44, 121

Soft Law, xi, 49-50, 61, 63

Stockholm Convention on Persistent Organic Pollutants, xi-14, 47, 49, 52, 55-56, 60, 62-63, 70-71, 77-78, 89-90, 98-103, 105, 107-108, 118-119, 122-125, 127

Stockpiles, 75, 87, 97, 100, 112, 134

Sweden, vii, 40, 45-47, 68-70, 87, 95, 100, 102, 118-119, 122

Swedish Environmental Code
 See Environmental Code

Swedish Government, 45-46, 60

Swedish Minister of the Environment, 45

Synthetic Organic Chemical Substances
 See Chemicals

Toxaphene, 2-3, 17-20, 22, 25, 27-30, 32, 34, 37, 39, 48, 61, 69, 79, 96, 103, 151

Transfer of Technology, 65, 108, 128, 140

UNCED
 See United Nations Conference on the Environment and Development

UNEP
 See United Nations Environmental Program

UNEP Governing Council Decision 18/31, 81

UNEP Governing Council Decision 18/32, 2, 59-60, 70, 79-80, 82

UNEP Governing Council Decision 19/13C, 80, 86, 127

UNIDO
 See United Nations Industrial Development Organization

Unintended Produced By-products and Contaminants, 34

Union des Advocats, 104

UNITAR
 See United Nations Institute for Training and Research

United Nations Conference on the Environment and Development, 50, 55-56, 104

United Nations Convention on the Law of the Sea
 See Law of the Sea

United Nations Economic and Social Council, 104-105

United Nations Environmental Program, xi, 5, 13, 17-20, 22, 25, 27-28, 30, 32, 34, 37, 50-51, 54-55, 57-60, 63-64, 68, 70-72, 75, 79-91, 93-103, 105, 118

United Nations Industrial Development Organization, 59, 85, 101

United Nations Institute for Training and Research, 59

Vienna Convention on the Law of Treaties, 63, 117

Washington Declaration on Protection of the Marine Environment from Land-Based Activities, 49, 82

WHO
 See World Health Organization

Withdrawal, 90, 119, 131, 150, 156

World Bank, 43, 55, 85, 88, 101, 115

World Council of Churches, 104

World Health Organization, xi, 10, 15-16, 22, 24, 59-60, 78-79, 84-85, 110, 155-156, 158

World Trade Organization, 90, 104

World Wildlife Fund, 84, 90, 104, 109

WTO supremacy clause, 90